Architecting Cloud-Native Serverless Solutions

Design, build, and operate serverless solutions on cloud and open source platforms

Safeer CM

BIRMINGHAM—MUMBAI

Architecting Cloud-Native Serverless Solutions

Group Product Manager: Preet Ahuja
Publishing Product Manager: Surbhi Suman
Content Development Editor: Sujata Tripathi
Technical Editor: Arjun Varma
Copy Editor: Safis Editing
Project Coordinator: Sean Lobo
Proofreader: Safis Editing
Indexer: Rekha Nair
Production Designer: Ponraj Dhandapani
Marketing Coordinator: Rohan Dobhal

First published: June 2023

Production reference: 2150623

Published by Packt Publishing Ltd.
Livery Place
35 Livery Street
Birmingham
B3 2PB, UK.

ISBN 978-1-80323-008-5

www.packtpub.com

This book is dedicated to my beloved family: my parents, Mohamed Shafi and Rukiyya, who granted me the freedom to make my own choices, my wife, Sani, whose unwavering support propels me forward, and my sons, Aman and Ayaan, whose energetic presence keeps me on my toes every day.

Contributors

About the author

Safeer CM has been working in site reliability, DevOps, and platform engineering for 17 years. He is passionate about systems, networks, and anything in between. His recent focus areas include serverless computing, cloud-native landscape, and multi-cloud infrastructure. He is also an SRE and DevOps practitioner and consultant. Based out of Bengaluru (Bangalore), India, Safeer works as a senior staff SRE for Flipkart, the Indian e-commerce leader. Safeer has written several articles and has been a speaker at and organizer of multiple meetups. He is currently an ambassador for the Continuous Delivery Foundation, where he helps the organization with community adoption and governance.

A heartfelt thanks to my family for all their support. This book would not have been possible without the countless mentors and well-wishers who imparted their knowledge and kindness, which kept me going.

About the reviewers

Kuldeep Singh is an experienced data center migration architect and leader. He has expertise in infrastructure project management and has delivered successful data center projects for telecom giants such as AT&T (US) and Vodafone (Europe). He also holds various certifications for cloud solutions and project management. In his free time, he likes to keep fit by lifting weights at the gym.

I am grateful to my wife, my kid Mozo, and my family, for their constant support and encouragement throughout this project. I also appreciate the publisher for their professionalism and guidance in the publishing process. I hope this book will be useful and enjoyable for the readers.

Aditya Krishnakumar has worked in DevOps for the past 5+ years, with 3 years specifically spent working with serverless platforms on AWS and Google Cloud. He is currently working as a senior site reliability engineer at a US-based product-based organization that provides a DevSecOps platform designed for applications hosted on Kubernetes and containers. He previously worked for **Boston Consulting Group** (**BCG**), where he was responsible for the operations of a big data analytics platform hosted in Google Cloud. He provides contributions to the DevOps community via his blog and is a member of AWS Community Builders, a global community of AWS and cloud enthusiasts.

I'd like to thank my family, who have believed in me and understood the time and commitment it takes to review and test the contents of a book. Working in this field would not have been possible without the support of the DevOps community that has developed over the past several years.

Shubham Mishra is a prominent figure in the field of ethical hacking and is widely recognized as India's youngest cybersecurity expert. He is the founder and CEO of TOAE Security Solutions, a company dedicated to the development of robust cybersecurity methods used worldwide. With over a decade of experience, Shubham has worked with some of the largest companies globally, providing updated and relevant content for the industry.

In addition to his professional achievements, Shubham is also an accomplished author, having written three books on various aspects of cybersecurity. His contributions to the field have cemented his reputation as a leading authority in the world of ethical hacking and cybersecurity.

Table of Contents

Part 2 – Platforms and Solutions in Action

3

Serverless Solutions in AWS   57

6

Serverless Cloudflare   151

7

Kubernetes, Knative and OpenFaaS   183

8

Self-Hosted FaaS with Apache OpenWhisk   217

Part 3 – Design, Build, and Operate Serverless

9

Implementing DevOps Practices for Serverless   243

Preface

I still remember the first time I learned about AWS Lambda, shortly after its general availability. My introduction to and interest in the serverless world was triggered by it. What started as a casual experiment of a niche cloud service soon turned into a journey of exploring serverless as a computing paradigm and how to fit more business use cases into it. As months and years went by, more vendors started rolling out their own **Function-as-a-Service** (**FaaS**) alternatives as well as other serverless backends. It was an amazing evolution that brought in a new era of cloud computing.

As exciting as it is, with the explosion of serverless technologies and vendors, it has become challenging to even comprehend the foundational services and what they have to offer. With the era of hybrid cloud, it is not necessary to stick to one cloud vendor and their serverless platforms. You can do your due diligence and mix and match the services from different vendors or even build a private serverless cloud. But to make those decisions, it is vital to know about the choices you have across the vendors and open source technologies of serverless. While there is extensive documentation that each of these vendors supply, there was no one place from where an engineer could understand the basics of serverless offerings across the spectrum. This book is written with that audience in mind.

This book lays a strong foundation for serverless landscape and technologies in a vendor-agnostic manner. You'll learn how to select the different cloud vendors and technology as per your needs. You'll deep dive into the serverless services across **Amazon Web Services** (**AWS**), **Google Cloud Platform** (**GCP**), Azure, and Cloudflare, followed by open source serverless tools such as Knative, OpenFaaS, and OpenWhisk, along with their examples. You'll explore serverless solutions on Kubernetes that can run on cloud-hosted clusters and on-premises with the help of real-world use cases. You'll be able to extend your learnings by understanding the development frameworks, DevOps approaches, best practices, security, and design principles of serverless.

By the end of this serverless book, you'll be equipped with the skills required to solve your business problems by using the appropriate serverless vendors and technologies and build efficient and cost-effective serverless systems on your own.

Who this book is for

This book is intended for platform, DevOps, site reliability engineers, and application developers who want to start building serverless solutions. It will be a valuable reference for solution architects trying to modernize a legacy application or working on a greenfield project. Besides, it is helpful for anyone solving business or operational problems without wanting to manage complicated technology infrastructure using serverless technologies.

What this book covers

Chapter 1, Serverless Computing and Function as a Service, introduces serverless technologies and the need for serverless. The chapter starts with an introduction to the evolution of the cloud and how internet infrastructure evolved over time. Then it dives into FaaS, its history, how FaaS works, and serverless in general.

Chapter 2, Backend as a Service and Powerful Serverless Platforms, talks about serverless backends and advanced backend services. While FaaS provides computing power in the serverless world, compute in isolation doesn't add business value. This is where serverless backends come into the picture. The chapter discusses various serverless backends, including object stores, message queues, workflow orchestrators, and so on.

Chapter 3, Serverless Solutions in AWS, covers serverless solutions in AWS, starting with its flagship FaaS service: Lambda. Then it covers the core serverless services such as **Simple Storage Service** (**S3**), **Simple Queue Service** (**SQS**), DynamoDB, and so on. The chapter also delves into other serverless backends such as **Simple Notification Service** (**SNS**), EventBridge, Step Functions, and Cognito.

Chapter 4, Serverless Solutions in Azure, covers the Microsoft Azure cloud's serverless offering. The most important service is the FaaS offering: Azure Functions. This chapter also covers other critical services such as Azure Cosmos DB, Logic Apps, Event Grid, Event Hubs, and so on.

Chapter 5, Serverless Solutions in GCP, covers the most important serverless services that Google Cloud offers. The chapter starts by covering in detail Google Cloud Functions, followed by the critical serverless backends such as Pub/Sub, Cloud Storage, Workflows, and so on. It also touches upon the newer serverless services such as Cloud Run, Eventarc, Cloud Scheduler, and many database services.

Chapter 6, Serverless Cloudflare, talks about how Cloudflare has a unique position among the serverless vendors by offering a number of services in its core role as a network and edge services vendor. Its FaaS platform – Cloudflare Workers – was brought in to supercharge its edge services. From that, Cloudflare's serverless portfolios have evolved to add more services such as Workers KV, Cloudflare Pages, R2, and so on. This chapter covers all these services with examples and code snippets.

Chapter 7, Kubernetes, Knative, and OpenFaaS, covers the most cloud-native serverless solutions – based on Kubernetes. Knative is a cloud-native FaaS implementation that can be deployed in any Kubernetes cluster, which also powers Google Cloud Functions behind the scenes. This chapter covers its implementation with examples using a Kubernetes cluster. OpenFaaS is another open source platform that can be run on Kubernetes or on virtual machines, and the chapter covers the architecture as well as usage with examples.

Chapter 8, Self-Hosted FaaS with Apache OpenWhisk, talks about Apache OpenWhisk, which is an open source FaaS platform hosted and maintained by the Apache Foundation. It powers the FaaS platforms for IBM Cloud, Adobe I/O Runtime, and so on. This chapter covers its architecture and implementation with examples.

Chapter 9, Implementing DevOps Practices for Serverless, covers multiple frameworks that are helpful to implement your serverless story. It starts with the coverage of "the Serverless Framework," followed by another framework called Zappa – both of which make serverless development easy. This chapter also talks about **Infrastructure as Code (IaC)** and its implementation with Terraform and Pulumi.

Chapter 10, Serverless Security, Observability, and Best Practices, talks about serverless security, which I find is a topic that is not given enough attention, where most of the security responsibility is offloaded to the vendor. This chapter examines the cloud responsibility matrix and the security best practices laid out by different organizations. It also talks about observability for serverless services and closes the chapter with industry best practices.

Chapter 11, Architectural and Design Patterns for Serverless, has the central theme of how to design better serverless solutions. To help with this, this chapter starts by giving an introduction to software design patterns followed by cloud architecture patterns. Then, this chapter discusses the Well-Architected Framework and patterns for serverless applications.

To get the most out of this book

Wherever possible, use the cloud shell provided by the cloud vendor for easier bootup. AWS, Azure, and GCP provide shells at the time of writing.

Software/hardware covered in the book	Operating system requirements
AWS, GCP, Azure, Cloudflare	Windows, macOS, or Linux
Kubernetes, Knative	A terminal or command-line application in your preferred operating system, plus a browser
OpenFaas, OpenWhisk	Knowledge of at least one cloud platform and the basics of Linux will help you
Serverless Framework, Zappa	
Terraform, Pulumi	

If you are using the digital version of this book, we advise you to type the code yourself or access the code from the book's GitHub repository (a link is available in the next section). Doing so will help you avoid any potential errors related to the copying and pasting of code.

Download the example code files

You can download the example code files for this book from GitHub at `https://github.com/PacktPublishing/Architecting-Cloud-Native-Serverless-Solutions`. If there's an update to the code, it will be updated in the GitHub repository.

We also have other code bundles from our rich catalog of books and videos available at https://github.com/PacktPublishing/. Check them out!

Download the color images

We also provide a PDF file that has color images of the screenshots and diagrams used in this book. You can download it here: https://packt.link/2fHDU.

Conventions used

There are a number of text conventions used throughout this book.

Code in text: Indicates code words in text, database table names, folder names, filenames, file extensions, pathnames, dummy URLs, user input, and Twitter handles. Here is an example: "The /whisk.system namespace is reserved for entities that are distributed with the OpenWhisk system."

A block of code is set as follows:

```
root@serverless101:~/hello-worker# cat index.js
addEventListener('fetch', event => {
  event.respondWith(handleRequest(event.request))
})
/**
 * Respond with hello worker text
 * @param {Request} request
 */
async function handleRequest(request) {
  return new Response('Hello worker!', {
    headers: { 'content-type': 'text/plain' },
  })
}
```

When we wish to draw your attention to a particular part of a code block, the relevant lines or items are set in bold:

```
root@serverless101:~/hello-worker# wrangler dev -i 0.0.0.0
Listening on http://0.0.0.0:8787
watching "./"
```

Any command-line input or output is written as follows:

```
aws cloudformation create-stack \
  --stack-name MyS3Bucket \
  --template-body file://tmp/cfntemplates/mys3.yml
```

Bold: Indicates a new term, an important word, or words that you see onscreen. For instance, words in menus or dialog boxes appear in **bold**. Here is an example: "The Kubernetes **hairpin-mode** shouldn't be **none** as OpenWhisk endpoints should be able to loop back to themselves."

> Tips or important notes
> Appear like this.

Get in touch

Feedback from our readers is always welcome.

General feedback: If you have questions about any aspect of this book, email us at customercare@packtpub.com and mention the book title in the subject of your message.

Errata: Although we have taken every care to ensure the accuracy of our content, mistakes do happen. If you have found a mistake in this book, we would be grateful if you would report this to us. Please visit www.packtpub.com/support/errata and fill in the form.

Piracy: If you come across any illegal copies of our works in any form on the internet, we would be grateful if you would provide us with the location address or website name. Please contact us at copyright@packt.com with a link to the material.

If you are interested in becoming an author: If there is a topic that you have expertise in and you are interested in either writing or contributing to a book, please visit authors.packtpub.com.

Share Your Thoughts

Once you've read *Architecting Cloud-Native Serverless Solutions*, we'd love to hear your thoughts! Scan the QR code below to go straight to the Amazon review page for this book and share your feedback.

https://packt.link/r/1803230088

Your review is important to us and the tech community and will help us make sure we're delivering excellent quality content.

Download a free PDF copy of this book

Thanks for purchasing this book!

Do you like to read on the go but are unable to carry your print books everywhere?

Is your eBook purchase not compatible with the device of your choice?

Don't worry, now with every Packt book you get a DRM-free PDF version of that book at no cost.

Read anywhere, any place, on any device. Search, copy, and paste code from your favorite technical books directly into your application.

The perks don't stop there, you can get exclusive access to discounts, newsletters, and great free content in your inbox daily

Follow these simple steps to get the benefits:

1. Scan the QR code or visit the link below

https://packt.link/free-ebook/9781803230085

2. Submit your proof of purchase

3. That's it! We'll send your free PDF and other benefits to your email directly

Part 1 – Serverless Essentials

This part provides the foundational knowledge on serverless computing. We will understand the history and evolution of serverless, what **Function as a Service (FaaS)** is, why there is more to serverless than FaaS, and so on. We will explore the most important serverless services provided by top vendors, as well as the self-hosted alternatives and the special cases in serverless computing.

This part has the following chapters:

- *Chapter 1, Serverless Computing and Function as a Service*
- *Chapter 2, Backend as a Service and Powerful Serverless Platforms*

1

Serverless Computing and Function as a Service

Serverless computing has ushered in a new era to an already revolutionizing world of cloud computing. What started as a nascent idea to run code more efficiently and modularly has grown into a powerful platform of serverless services that can replace traditional microservices and data pipelines in their entirety. Such growth and adoption also brings in new challenges regarding integration, security, and scaling. Vendors are releasing newer services and feature additions to existing services all around, opening more and more choices for customers.

AWS has been a front runner in serverless offerings, but other vendors are catching up fast. Replacing in-house and self-hosted applications with serverless platforms is becoming a trend. **Function as a Service (FaaS)** is what drives serverless computing. While all cloud vendors are offering their version of FaaS, we are also seeing the rise of self-hosted FaaS platforms, making this a trend across cloud and data center infrastructures alike. People are building solutions that are cloud agnostic using these self-hosted platforms as well.

In this chapter, we will cover the foundations of serverless and FaaS computing models. We will also discuss the architecture patterns that are essential to serverless models.

In this chapter, we will cover the following topics:

- Evolution of computing in the cloud

- Serverless and FaaS

- Microservice architecture

- Event-driven architecture

- FaaS in detail

- API gateways and the rise of serverless APIs

- The case for serverless

Evolution of computing in the cloud

In this section, we will touch on the evolution of cloud computing and why the cloud matters. We will briefly cover the technologies that drive the cloud and various delivery models.

Benefits of cloud computing

Cloud computing has revolutionized IT and has spearheaded unprecedented growth in the past decade. By definition, cloud computing is the availability and process of delivering computing resources on-demand over the internet. The traditional computing model required software services to invest heavily in the computing infrastructure. Typically, this meant renting infrastructure in a data center – usually called colocation – for recurring charges per server and every other piece of hardware, software, and internet they used. Depending on the server count and configurations, this number would be pretty high and was inflexible in the billing model – with upfront costs and commitments. If more customized infrastructure with access to network gears and more dedicated internet bandwidth is required, the cost would go even higher and it would have more upfront costs and commitments. Internet-scale companies had to build or rent entire data centers across the globe to scale their applications – most of them still do.

This traditional IT model always led to a higher total cost of ownership, as well as higher maintenance costs. But these were not the only disadvantages – lack of control, limited choices of hardware and software combinations, inflexibility, and slow provisioning that couldn't match the market growth and ever-increasing customer bases were all hindering the speed of delivery and the growth of applications and services. Cloud computing changed all that. Resources that were available only by building or renting a data center were now available over the internet, at a click of a button or a command. This wasn't just the case servers, but private networks, routers, firewalls, and even software services and distributed systems – which would take traditional IT a huge amount of manpower and money to maintain – were all available right around the virtual corner.

Cost has always been a crucial factor in deciding on which computing model to use and what investment companies are willing to make in the short and long term. In the next section, we will talk about the difference between the cost models in the cloud.

CAPEX versus OPEX

The impact of cloud computing is multifold. On one hand, it allows engineering and product teams to experiment with their products freely without worrying about planning for the infrastructure quarters or even years back. It also has the added benefit of not having to actively manage the cloud resources, unlike the data center infrastructure. Another reason for its wider adoption is the cost factor. The difference between traditional IT and the cloud in terms of cost is sometimes referred to as **CAPEX** versus **OPEX**.

CAPEX, also known as capital expenditure, is the initial and ongoing investments that are made in assets – IT infrastructure, in this case – to reap the benefits for the foreseeable future. This also includes the ongoing maintenance cost as it improves and increases the lifespan of the assets. On the other hand, the cloud doesn't require you to invest upfront in assets; the infrastructure is elastic and virtually unlimited as far as the customer is concerned. There is no need to plan for infrastructure capacity months in advance, or even worry about the underutilization of already acquired IT assets. Infrastructure can be built, scaled up or down, and ultimately torn down without any cost implications. The expenditure, in this case, is operating expenditure – OPEX. This is the cost that's incurred in running the day-to-day business and what's spent on utilities and consumables rather than long-term assets. The flexible nature of cloud assets makes them consumable rather than assets.

Let's look at a few technologies that accelerated the adoption of the cloud.

Virtualization, software-defined networking, and containers

While we understand and appreciate cloud computing and the benefits it brings, the technologies that made it possible to move from traditional data centers to the cloud need to be acknowledged.

The core technology that succeeded in capitalizing on the potential of hardware and building abstraction on top of it was virtualization. It allowed virtual machines to be created on top of the hardware and the host operating system. Network virtualization soon followed, in the form of **Software-Defined Networking (SDN)**. This allowed vendors to provide a completely virtualized private network and servers on top of their IT infrastructure. Virtualization was prevalent much before cloud computing started but was limited to running in data centers and development environments, where the customers or vendors directly managed the entire stack, from hardware to applications.

The next phase of technological revolution came in the form of containers, spearheaded by Docker's container runtime. This allowed process, network, and filesystem isolation from the underlying operating system. It was also possible to enforce resource utilization limits on the processes running inside the container. This amazing feat was powered by Linux namespaces, cgroups, and Union Filesystem. Packaging runtimes and application code into containers led to the dual benefit of portability and a lean operating system. It was a win for both application developers and infrastructure operators.

Now that you are aware of how virtualization, SDN, and containers came around, let's start exploring the different types of cloud computing.

Types of cloud computing

In this section, we are going to look at different cloud computing models and how they differ from each other.

Public cloud

The public cloud is the cloud infrastructure that's available over the public internet and is built and operated by cloud providers such as Amazon, Azure, Google, IBM, and so on. This is the most common cloud computing model and is where the vendor manages all the infrastructure and ensures there's enough capacity for all use cases.

A public cloud customer could be anyone who signs up for an account and has a valid payment method. This provides an easy path to start building on cloud services. The underlying infrastructure is shared by all the customers of the public cloud across the globe. The cloud vendor abstracts out this shared-ness and gives each customer the feeling that they have a dedicated infrastructure to themselves. The capacity is virtually unlimited, and the reliability of the infrastructure is guaranteed by the vendor. While it provides all these benefits, the public cloud can also cause security loopholes and an increased attack surface if it's not maintained well. Excessive billing can happen due to a lack of knowledge of the cloud cost model, unrealistic capacity planning, or abandoning the rarely used resources without disposing of them properly.

Private cloud

Unlike with the public cloud, a private cloud customer is usually a single business or organization. A private cloud could be maintained in-house or in the company-owned data centers – usually called internal private clouds. Some third-party providers run dedicated private clouds for business customers. This model is called a hosted private cloud.

A private cloud provides more control and customization for businesses, and certain businesses prefer private clouds due to their business nature. For example, telecom companies prefer to run open source-based private clouds – Apache OpenStack is the primary choice of technology for a large number of telecom carriers. Hosting the cloud infrastructure also helps them integrate the telco hardware and network with the computing infrastructure, thereby improving their ability to provide better communication services. This added flexibility and control also comes at a cost – the cost of operating and scaling the cloud. From budget planning to growth predictions, to hardware and real estate acquisition for expansion, this becomes the responsibility of the business. The engineering cost – both in terms of technology and manpower – becomes a core cost center for the business.

Hybrid cloud

The hybrid cloud combines a public cloud and a physical infrastructure – either operated on-premises or on a private cloud. Data and applications can move between the public and private clouds securely to suit the business needs. Organizations could adopt a hybrid model for many reasons; they could be bound by regulations and compliance (such as financial institutions), low latency for certain applications to be placed close to the company infrastructure, or just because huge investments have already been made in the physical infrastructure. Most public clouds identify this as a valid business use case and provide cloud solutions that offer connectivity from cloud infrastructure to data centers through a private WAN-wide area network. Examples include AWS Direct Connect, GCP Interconnect, and Azure ExpressRoute.

An alternate form of hybrid cloud is the multi-cloud infrastructure. In these scenarios, one public cloud infrastructure is connected to one or more cloud infrastructures hosted by different vendors:

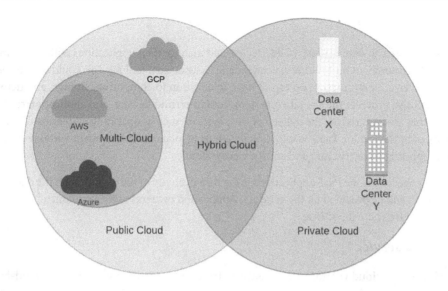

Figure 1.1 – Types of cloud computing

The preceding diagram summarizes the cloud computing types and how they are interrelated. Now that we understand these types, let's look at various ways in which cloud services are delivered.

Cloud service delivery models – IaaS, PaaS, and SaaS

While cloud computing initially started with services such as computing and storage, it soon evolved to offer a lot more services that handle data, computing, and software. These services are broadly categorized into three types based on their delivery models: **Infrastructure as a Service (IaaS)**, **Platform as a Service (PaaS)**, and **Software as a Service (SaaS)**. Let's take a quick look at each of these categories.

Infrastructure as a service

In IaaS, the cloud vendor delivers services such as compute (virtual machines, containers, and so on), storage, and network as a cloud service – just like a traditional data center would. It also covers a lot of supporting services, such as firewall and security, monitoring, load balancing, and more. Out of all the service categories listed, IaaS provides the most control to the customer and they get to fine-tune and configure core services, as they would in a traditional IT infrastructure.

While the compute, storage, and network are made available to the customers as infrastructure pieces, these are not actual physical hardware. Instead, these resources are virtualized – as abstractions on top of the real hardware. There is a lesser-known variant of IaaS where the real hardware is directly provisioned and exposed to the customer. This category of services is called **Bare-Metal as a Service (BMaaS)**. BMaaS provides much more control than IaaS to the customer and it is also usually costlier and takes more engineering time to manage.

Platform as a service

PaaS allows customers to develop, test, build, deploy, and manage their applications without having to worry about the resources or the build environment and its associated tooling. This could be considered as an additional layer of abstraction on top of IaaS. In addition to computing, storage, and network resources, PaaS also provides the operating system, container/middleware, and application runtime. Any updates that are required for the upkeep of the platform, such as operating system patching, will be taken care of by the vendor. PaaS enables organizations to focus on development without worrying about the supporting infrastructure and software ecosystem.

Any data that's needed for the PaaS applications is the responsibility of the user, though the data stores that are required will be provided by the vendors. Application owners have direct control of the data and can move it elsewhere if necessary.

Software as a service

In the SaaS model, a cloud vendor provides access to software via the internet. This cloud-based software is usually provided through a pay-as-you-go model where different sets of features of the same software are offered for varying charges. The more features used, the costlier the SaaS is. The pricing models also depend on the number of users using the software.

The advantage of SaaS is that it completely frees a customer from having to develop or operate their software. All the hassle of running such an infrastructure, including security and scaling, are taken care of by the vendor. The only commitment from the customer is the subscription fee that they need to pay. This freedom also comes at the cost of complete vendor dependency. The data is managed by the vendor; most vendors would enable their customers to take a backup of their data since finding a compatible vendor or reusing that data in-house could become challenging:

Data Center	IaaS	PaaS	SaaS
Application	Application	Application	Application
Data	Data	Data	Data
Runtime	Runtime	Runtime	Runtime
Middleware	Middleware	Middleware	Middleware
Operating System	Operating System	Operating System	Operating System
Virtualization and SDN	Virtualization and SDN	Virtualization and SDN	Virtualization and SDN
Bare-Metal Servers	Bare-Metal Servers	Bare-Metal Servers	Bare-Metal Servers
Storage	Storage	Storage	Storage
Network	Network	Network	Network

Customer Managed	Vendor Managed

Figure 1.2 – Cloud service delivery models

Now that we have cemented our foundations of cloud computing, let's look at a new model of computing – *FaaS*.

Serverless and FaaS

In the previous sections, we discussed various types of clouds, cloud service delivery models, and the core technologies that drove this technology revolution. Now that we have established the baselines, it is time to define the core concept of this book – serverless.

When we say *serverless*, what we are usually referring to is an application that's built on top of a serverless platform. Serverless started as a new cloud service delivery model where everything except the code is abstracted away from the application developer. This sounds like PaaS as there are no servers to manage and the application developer's responsibility is limited to writing the code. There are some overlaps, but there are a few distinctive differences between PaaS and serverless, as follows:

PaaS	Serverless
Always-on application	Runs on demand
Scaling requires configuration	Automatic scaling
More control over the development and deployment infrastructure	Very limited control over the development and deployment infrastructure
High chance of idle capacity	Full utilization and no idle time, as well as visibility to fine-tune and benchmark business logic
Billed for the entirety of the application's running time	Billed every time the business logic is executed

Table 1.1 – PaaS versus serverless

In the spectrum of cloud service delivery models, serverless can be placed between PaaS and SaaS.

FaaS and BaaS

The serverless model became popular in 2014 after AWS introduced a service called **Lambda**, which provides FaaS. Historically, other services could be considered ancestors of serverless, such as Google App Engine and `iron.io`. Lambda, in its initial days, allowed users to write functions in a selected set of language runtimes. This function could then be executed in response to a limited set of events or be scheduled to run at an interval, similar to a cronjob. It was also possible to invoke the function manually.

As we mentioned previously, Lambda was one of the first services in the category of FaaS and established itself as a standard. So, when we say serverless, people think of FaaS and, subsequently, Lambda. But FaaS is just one part of the puzzle – it serves as the computing component of serverless. As is often the case, compute is meaningless without data and a way to provide input and output. This is where a whole range of supporting services come into the picture. There are services in the category of API gateways, object stores, relational databases, NoSQL databases, communication buses, workflow management, authentication services, and more. In general, these services power the backend for serverless computing. These services can be categorized as **Backend as a Service (BaaS)**. We will look at BaaS in the next chapter.

Before we get into the details of FaaS, let's review two architecture patterns that you should know about to understand serverless – the microservice architecture and the **Event-Driven Architecture (EDA)**.

Microservice architecture

Before we look at the microservice architecture, let's look at how web applications were built before that. The traditional way of building software applications was called monolithic architecture. Enterprises used to develop applications as one big indivisible unit that provided all the intended functionality. In the initial phases of development and deployment, monoliths offered some fairly good advantages. Project planning and building a minimum viable product – the alpha or beta version – was easier. A single technology stack would be chosen, which made it easier to hire and train developers. In terms of deployment, it was easier to scale since multiple copies of this single unit could be thrown behind a load balancer to scale for increased traffic:

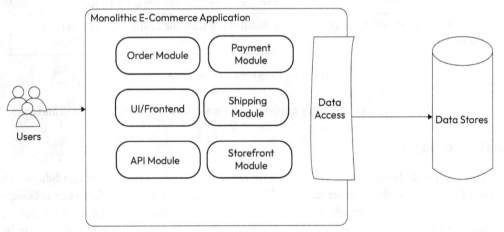

Figure 1.3 – Monolithic architecture

The problem starts when the monolithic application has to accommodate more features and the business requirements grow. It becomes increasingly complex to understand the business logic and how the various pieces that implement the features are interconnected. As the development team grows, parts of the application will be developed by dedicated teams. This will lead to a disconnect in communication and introduce non-compatible changes and more complex dependencies. The adoption of new technologies will become virtually impossible and the only choice to bring in changes that align with changing business requirements would be to rewrite the application in its entirety. On the scaling front, the problem is that we need to scale up the entire application, even if only a particular piece of code or business logic is creating the bottleneck. This inflexibility causes unnecessary provisioning of resources and idle time when the particular business logic is not in the critical path.

The microservice architecture was introduced to fix the shortcomings of the monolithic architecture. In this architecture, an application is organized as a collection of smaller independent units called microservices. This is achieved by building separate services around independent functions or the business logic of the application. In a monolithic architecture, the different modules of the application would communicate with each other using library calls or inter-process communication channels. In the case of the microservice architecture, individual services communicate with each other via APIs using protocols such as HTTP or gRPC. Some of the key features of the microservice model are as follows:

- Loosely coupled – each unit is independent.
- Single responsibility – one service is responsible for one business function.
- Independently develop and deploy.
- Each service can be built in a separate technology stack.
- Easier to divide and separate the backends that support the services, such as databases.
- Smaller and separate teams are responsible for one or more microservices.
- The developer's responsibilities are better and clearly defined.
- Easy to scale independently.

- A bug in one service won't bring down the entire application. Instead, a single piece of business logic or a feature would be impacted:

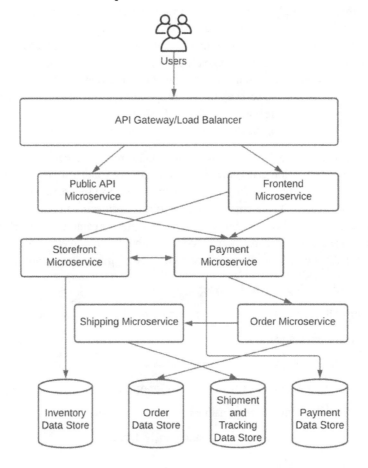

Figure 1.4 – E-commerce application with the microservice architecture

While microservices help solve a lot of problems that the monolithic architecture posed, it is no silver bullet. Some of the disadvantages of microservices are as follows:

- Given all inter-microservice communication happens via the network, network issues such as latency have a direct impact and increase the time it takes to communicate between two parts of the business function.

- Since most business logic requires talking to other microservices, it increases the complexity of managing the service.

- Debugging becomes hard in a distributed microservice environment.

- More external services are required to ensure visibility into the infrastructure using metrics, logs, and tracing. The absence of any of this makes troubleshooting hard.

- It puts a premium on monitoring and increases the overall infrastructure cost.

- Testing global business logic would involve multiple service calls and dependencies, making it very challenging.

- Deployments require more standardization, engineering investment, and continuous upkeep.

- It's complex to route requests.

This sums up the microservice architecture and its benefits. In the next section, we will briefly discuss a few technologies that can help microservices be deployed more structurally.

Containers, orchestration, and microservices

Containers revolutionized the way we deploy and utilize system resources. While we had microservices long before containers became popular, they were not configured and deployed optimally. A container's capability to isolate running processes from one another and limit the resources that are used by processes was a great enabler for microservices. The introduction of container orchestration services such as Kubernetes took this to the next level. It helped support more streamlined deployments and developers could define every resource, every network, and every backend for an application using a declarative model. Currently, containers and container orchestration are the de facto way to deploy microservices.

Now that we have a firm understanding of the microservice architecture, let's examine another architecture pattern – EDA.

Event-driven architecture

EDA is an architectural pattern where capturing, processing, and storing events is the central theme. This allows a bunch of microservices to exchange and process information asynchronously. But before we dive into the details of the architecture, let's define what an event is.

Events

An event is the record of a significant occurrence or change that's been made to the state of a system. The source of the event could be a change in the hardware or software system. An event could also be a change to the content of a data item or a change in the state of a business transaction. Anything that happens in your business or IT infrastructure could be an event. Which events do we need to process and bring under EDA as an engineering and business choice? Events are immutable records and can be read and processed without the event needing to be modified. Events are usually ordered based on their creation time.

Some examples of events are as follows:

- Customer requests
- Change of balance in a bank account
- A food delivery order being placed
- A user being added to a server
- Sensor reading from a hardware or IoT device
- A security breach in a system

You can find examples of events all around your application and infrastructure. The trick is deciding on which are relevant and need processing. In the next section, we'll look at the structure of EDA.

Structure and components of an EDA

The value proposition of EDA comes from the fact that an event loses its processing value as it gets older. Event-driven systems can respond to such events as they are generated and take appropriate action to add a lot of business value. In an event-driven system, messages from various sources are ingested, then sent to interested parties (read microservices) for processing, and then persisted to disk for a defined period.

EDA fundamentally differs from the synchronous model that's followed by APIs and web stacks, where a response must be returned for every request synchronously. This could be compared to a customer support center using phone calls versus emails to respond to customer requests. While phone calls take a lot of time and need the support agent to be manually responding to the request, the same time can be spent asynchronously replying to a bunch of emails, often with the help of automation. The same principle applies to request-response versus event-driven models. But just like this example, EDA is not a silver bullet and can't be used on all occasions. The trick is in finding the right use case and building on it. Most critical systems and customer-facing services still have to rely on the synchronous request-response model.

The components of an event-driven model can be broadly classified into three types – event producer, event router (broker), and event consumer. The event producers are one or more microservices that produce interesting events and post them to the broker. The event broker is the central component of this architecture and enables loose coupling between producers and consumers. It is responsible for receiving the events, serializing or deserializing them, if necessary, notifying the consumers of the new event, and storing them. Certain brokers also filter the events based on conditions or rules. The consumers can then consume the interesting events at their pace:

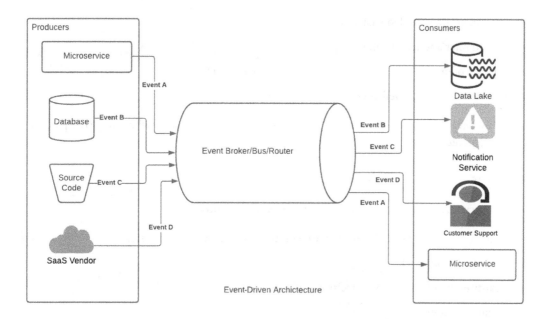

Figure 1.5 – EDA

That sums up the EDA pattern. Now, let's look into the benefits of EDA.

Benefits of EDA

The following is not a comprehensive list of the benefits of EDA, but this should give you a fair idea of why this architecture pattern is important:

- Improved scalability and fault tolerance due to a producer or consumer failing doesn't impact the rest of the systems.

- Real-time data processing for better decisions and customer experience – businesses can respond in real time to changes in customer behavior and make decisions or share data that improves the quality of the service.

- Operational stability and agility.

- Cost efficiency compared to batch processing. With batch processing, large volumes of data had to be stored and processed in batches. This meant allocating a lot more storage and compute resources for a longer period. Once batch processing is over, the computing resource becomes idle. This doesn't happen in EDA as the events are processed as they arrive, and it distributes the compute and storage optimally.

- Better interoperability between independent services.

- High throughput and low latency.

- Easy to filter and transform events.

- The rate of production and consumption doesn't have to match.

- Works with small as well as complex applications.

Now that we have covered the use cases of EDA, let's look at some use cases where the EDA pattern can be implemented.

Use cases

EDA has a very varied set of use cases; some examples are as follows:

- Real-time monitoring and alerting based on the events in a software system

- Website activity tracking

- Real-time trend analysis and decision making

- Fraud detection

- Data replication between similar and different applications

- Integration with external vendors and services

While EDA becomes more and more important as the business logic and infrastructure becomes complicated, there are certain downsides we need to be aware of. We'll explore them in the next section.

Disadvantages

As we mentioned earlier, EDA is no silver bullet and doesn't work with all business use cases. Some of its notable disadvantages are as follows:

- The decoupled nature of events can also make it difficult to debug or trace back the issues with events.

- The reliability of the system depends on the reliability of the broker. Ideally, the broker should be either a cloud service or a self-hosted distributed system with a high degree of reliability.

- Consumer patterns can make it difficult to do efficient capacity planning. If many of the consumers are services that wake up only at a defined interval and process the events, this could create an imbalance in the capacity for that period.

- There is no single standard in implementing brokers – knowing the guarantees that are provided by the broker is important. Architectural choices such as whether it provides a strong guarantee of ordering or the promise of no duplicate events should be figured out early in the design, and the producers and consumers should be designed accordingly.

In the next section, we will discuss what our software choices are for EDA, both on-premises and in the cloud.

Brokers

There are open source brokers such as Kafka, Apache Pulsar, and Apache ActiveMQ that can implement some form of message broker. Since we are mostly talking in the context of the cloud in this book, the following are the most common cloud brokers:

- Amazon **Simple Queue Service (SQS)**

- Amazon **Simple Notification Service (SNS)**

- Amazon EventBridge

- Azure Service Bus queues

- Azure Service Bus topics

- Google Cloud Pub/Sub

- Google Cloud Pub/Sub Lite

EDA, as we've discovered, is fundamental to a lot of modern applications' architectures. Now, let's look at FaaS platforms in detail.

FaaS in detail – self-hosted FaaS

We briefly discussed FaaS earlier. As a serverless computing service, it is the foundational service for any serverless stack. So, what exactly defines a FaaS and its *functionality*?

As in the general definition of a function, it is a discrete piece of code that can execute one task. In the context of a larger web application microservice, this function would ideally serve a single URL endpoint for a specific HTTP method – say, GET, POST, PUT, or DELETE. In the context of EDA, a FaaS function would handle consuming one type of event or transforming and fanning out the event to multiple other functions. In scheduled execution mode, the function could be cleaning up some logs or changing some configurations. Irrespective of the model where it is used, FaaS has a simple objective – to run a function with a set of resource constraints within a time limit. The function could be triggered by an event or a schedule or even manually launched.

Similar to writing functions in any language, you can write multiple functions and libraries that can then be invoked within the primary function code. So long as you provide a function to FaaS, it doesn't care about what other functions you have defined or libraries you have included within the code snippet. FaaS considers this function as the *handler* function – the name could be different for different platforms, but essentially, this function is the entry point to your code and could take arguments that are passed by the platform, such as an event in an event-driven model.

FaaS runtimes are determined and locked by the vendor. They usually decide whether a language is supported and, if so, which versions of the language runtime will be available. This is usually a limited list where each platform adds support for more languages every day. Almost all platforms support a minimum of Java, JavaScript, and Python.

The process to create and maintain these functions is similar across platforms:

- The customer creates a function, names it, and decides on the language runtime to use.

- The customer decides on the limit for the supported resource constraints. This includes the upper limit of RAM and the running time that the function will use.

- While different platforms provide different configuration features, most platforms provide a host of configurations, including logging, security, and, most importantly, the mechanism to trigger the function.

- All FaaS platforms support events, cron jobs, and manual triggers.

- The platform also provides options to upload and maintain the code and its associated dependencies. Most also support various versions of the function to be kept for rollbacks or to roll forward. In most cloud platforms, these functions can also be tested with dummy inputs provided by the customer.

The implementation details differ across platforms but behind the scenes, how FaaS infrastructure logically works is roughly the same everywhere. When a function is triggered, the following happens:

- Depending on the language runtime that's been configured for the function, a container that's baked with the language runtime is spun up in the cloud provider's infrastructure.

- The code artifact – the function code and dependencies that are packed together as an archive or a file – is downloaded from the artifact store and dropped into the container.

- Depending on the language, the command that's running inside the container will vary. But this will ultimately be the runtime that's invoking the *entry point* function from the artifact.

- Depending on the platform and how it's invoked, the application that's running in the container will receive an event or custom environment variables that can be passed into the entry point function as arguments.

- The container and the server will have network and access restrictions based on the security policy that's been configured for the function:

FaaS Infrastructure

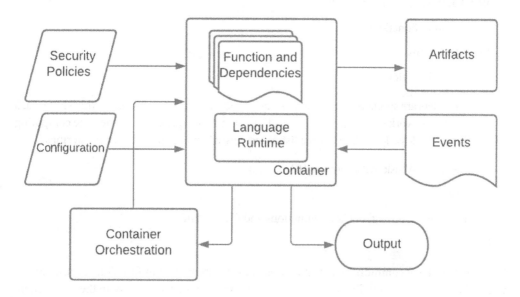

Figure 1.6 – FaaS infrastructure

One thing that characterizes FaaS is its stateless nature. Each invocation of the same function is an independent execution, and no context or global variables can be passed around between them. The FaaS platform has no visibility into the kind of business logic the code is executing or the data that's being processed. While this may look like a limiting factor, it's quite the opposite. This enables FaaS to independently scale multiple instances of the same function without worrying about the communication between them. This makes it a very scalable platform. Any data persistence that's necessary for the business logic to work should be saved to an external data service, such as a queue or database.

Cloud FaaS versus self-hosted FaaS

While FaaS started with the hosted model, the stateless and lightweight nature of it was very appealing. As it happens with most services like this, the open source community and various vendors created open source FaaS platforms that can be run on any platform that offers virtual machines or bare metal computing. These are known as self-hosted FaaS platforms. With self-hosted FaaS platforms, the infrastructure is not abstracted out anymore. Somebody in the organization will end up maintaining the infrastructure. But the advantage is that the developers have more control over the infrastructure and the infrastructure is much more secure and customizable.

The following is a list of FaaS offerings from the top cloud providers:

- AWS Lambda

- Google Cloud Functions

- Azure Functions

- IBM Cloud Functions

Other cloud providers are specialized in certain use cases, such as Cloudflare Workers, which is the FaaS from the edge and network service provider. This FaaS offering mostly caters to the edge computing use case within serverless. The following is a list of self-hosted and open source FaaS offerings:

- Apache OpensWhisk – also powers IBM Cloud

- Kubeless

- Knative – powers Google's Cloud Functions and Cloud Run

- OpenFaaS

All FaaS offerings have common basic features, such as the ability to run functions in response to events or scheduled invocations. But a lot of other features vary between platforms. In the next section, we will look at a very common serverless pattern that makes use of FaaS.

API gateways and the rise of serverless API services

API Gateway is an architectural pattern that is often part of an API management platform. API life cycle management involves designing and publishing APIs and provides tools to document and analyze them. API management enables enterprises to manage their API usage, respond to market changes quickly, use external APIs effectively, and even monetize their APIs. While a detailed discussion on API management is outside the scope of this book, one component of the API management ecosystem is of particular interest to us: API gateways.

An API gateway can be considered as a gatekeeper for all the API endpoints of the enterprise. A bare-bones API gateway would support defining APIs, routing them to the correct backend destination, and enforcing authentication and authorization as a minimum set of features. Collecting metrics at the API endpoints is also a commonly supported feature that helps in understanding the telemetry of each API. While cloud API gateways provide this as part of their cloud monitoring solutions, self-hosted API gateways usually have plugins to export metrics to standard metric collection systems or metric endpoints where external tools can scrape metrics. API gateways either host the APIs on their own or send the traffic to internal microservices, thus acting as API proxies. The clients of API gateways could be mobile and web applications, third-party services, and partner services. Some of the most common features of API gateways are as follows:

- **Authentication and authorization**: Most cloud-native API gateways support their own **Identity and Access Management** (**IAM**) systems as one of their leading authentication and authorization solutions. But as APIs, they also need to support common access methods using API keys, JWTs, mutual-TLS, and so on.

- **Rate limiting, quotas, and security**: Controlling the number of requests and preventing abuse is a common requirement. Cloud API gateways often achieve this by integrating with their CDN/global load balancers and DDoS protection systems.

- **Protocol translation**: Converting requests and responses between various API protocols, such as REST, WebSocket, GraphQL, and gRPC.

- **Load balancing**: With the cloud, this is a given as API Gateway is a managed service. For self-hosted or open source gateways, load balancing may need additional services or configuration.

- **Custom code execution**: This enables developers to modify requests or responses before they are passed down to downstream APIs or upstream customers.

Since API gateways act as the single entry point for all the APIs in an enterprise, they support various types of endpoint types. While most common APIs are written as REST services and use the HTTP protocol, there are also WebSocket, gRPC, and GraphQL-based APIs. Not all platforms support all of these protocols/endpoint types.

While API gateways existed independent of the cloud and serverless, they got more traction once cloud providers started integrating their serverless platforms with API Gateway. As in the case of most cloud service releases, AWS was the first to do this. Lambda was initially released as a private preview in 2014. In June 2015, 3 months after Lambda became generally available, AWS released API Gateway and started supporting integration with Lambda. Other vendors followed suit soon after. Due to this, serverless APIs became mainstream.

The idea of a serverless API is very simple. First, you must define an API endpoint in the supported endpoint protocol; that is, REST/gRPC/WebSocket/GraphQL. For example, in an HTTP-based REST API, this definition would include a URL path and an associated HTTP method, such as GET/POST/ PUT/DELETE. Once the endpoint has been defined, you must associate a FaaS function with it. When a client request hits said endpoint, the request and its execution context are passed to the function, which will process the request and return a response. The gateway passes back the response in the appropriate protocol:

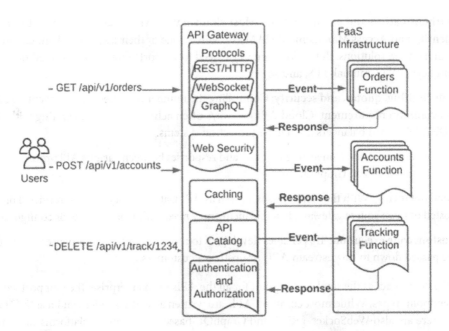

Figure 1.7 – API Gateway with FaaS

The advantage of serverless APIs is that they create on-demand APIs that can be scaled very fast and without any practical limits. The cloud providers would impose certain limits to avoid abuse and plan for better scalability and resource utilization of their infrastructure. But in most cases, you can increase these limits or lift them altogether by contacting your cloud vendor. In *Part 2* of this book, we will explore these vendor-specific gateways in detail.

The case for serverless

Serverless brought a paradigm shift in how infrastructure management can be simplified or reduced to near zero. However, this doesn't mean that there are no servers, but it abstracts out all management responsibilities from the customer. When you're delivering software, infrastructure management and maintenance is always an ongoing engineering cost and adds up to the operational cost – not to mention the engineering cost of having people manage the infrastructure for it. The ability to build lightweight microservices, on-demand APIs, and serverless event processing pipelines has a huge impact on the overall engineering cost and feature rollouts.

One thing we haven't talked about much is the cost model of FaaS. While only part of the serverless landscape, its billing model is a testament to the true nature of serverless. All cloud vendors charge for FaaS based on the memory and execution time the function takes for a single run. When used with precision, this cost model can shave off a lot of money from your cloud budget. Right-sizing the function and optimizing its code becomes a necessary skill for developers and will lead to a design-for-performance-first mindset.

As we will see in *Part 2* of this book, cloud vendors are heavily investing in building and providing serverless services as demand grows. The wide array of BaaS category services that are available to us is astounding and opens up a lot of possibilities. While not all business use cases can be converted to serverless, a large chunk of business use cases will find a perfect match in serverless.

Summary

In this chapter, we covered the foundations of serverless in general and FaaS in particular. How the microservice architecture has modernized the software architecture and how the latest container and container orchestration technologies are spearheading the microservice adoption was a key lesson. We also covered EDA and the API Gateway architecture. These concepts should have helped cement your foundational knowledge of serverless computing and will be useful when we start covering FaaS platforms in *Part 2* of this book. Serverless has evolved into a vast technology platform that encompasses many backends and computing services. The features and offerings may vary slightly between platforms and vendors, but the idea has caught up.

In the next chapter, we will look at some of the backend architectural patterns and technologies that will come in handy in serverless architectures.

2

Backend as a Service and Powerful Serverless Platforms

In the previous chapter, we covered the fundamentals of serverless, important architectural patterns, and **Function as a Service (FaaS)**. In this chapter, we are going to talk about the computing and data systems that can power up serverless. We will first talk about **Backend as a Service (BaaS)** and **Mobile BaaS (mBaaS)**, and then discuss a few technologies that are vital for a serverless architecture. The approach we take in discussing these technologies is to first introduce the concept, then talk about the **open source software (OSS)** that implements those technologies, and finally cover the corresponding cloud services. The reason for this approach is that we need to understand the fundamental building blocks conceptually, understand what would be the software you can use to implement these services in a cloud-agnostic way, and then understand the corresponding cloud solutions. This would help in implementing hybrid solutions that involve serverless and server-based services or migrating a legacy system partially or fully to a serverless stack. Besides the more sophisticated and complicated business problems you have to solve, you will end up using the best of both worlds. A purely serverless solution is neither ideal nor practical in such scenarios.

The important topics we are covering in this chapter are listed here:

- BaaS

- Messaging systems

- Object stores

- Workflow automation

- NoSQL platforms

- Edge computing

- **Internet of Things (IoT)** clouds

- Stream processing

- Future of serverless

We can divide the architecture of almost all software applications into frontend and backend technologies. Frontends deal with user interactions, while backends deal with everything else, including business logic, data stores, and other supporting services. Usually, the user could be a human who would use either a mobile application or a web application, but there is also the case of other software systems being the users, where one service would push or pull information from a system via **application programming interfaces (APIs)**. Here, the frontend application would use a **software development kit (SDK)** as the mechanism to interact with the APIs. In short, backends are any service that powers the user interaction offered by the frontends.

BaaS

BaaS is a computing model where BaaS acts as a middleware for mobile and web applications and exposes a unified interface to deal with one or more backend services. Since most of these services were focused on providing backend services for mobile application frontends, this was also called mBaaS. A prominent player in this arena is **Google Firebase**.

Firebase began as a start-up providing chat backend to mobile applications in 2011 and later evolved to providing real-time databases and authentication. In 2014, Google acquired the company and started integrating more tools into its portfolio. Many mobile and application development services from start-ups that Google acquired were later rolled into the Firebase portfolio. Many Google Cloud services such as Filestore, Messaging, and Push Notifications were integrated with Firebase later. Today, the integration with Google Cloud has become even more evolved, and a lot of services offered by Firebase can be managed from the Google Cloud Console and vice versa. As of today, Firebase provides a host of services broadly classified into three: Build, Release & Monitor, and (Customer) Engagement. You can view more information about these services in the following table:

Build	Release & Monitor	Engagement
Cloud Firestore	Google Analytics	Predictions
Realtime Database	Performance Monitoring	A/B Testing
Remote Config	Test Lab	In-App Messaging
Cloud Functions	App Distribution	Dynamic Links
Firebase ML	Crashlytics	
Authentication		
Cloud Messaging		
Cloud Storage		
Hosting		

Table 2.1 – Firebase services

I used Firebase here to show by example how it has evolved as a BaaS leader and the value it adds. To make a general definition, BaaS is one or more services provided for mobile and web application developers so that they can deliver quality applications without worrying about building complicated backend services. BaaS vendors try to anticipate most backend use cases an application developer might need and try to tailor their solutions accordingly. Some of the most common services that BaaS/mBaaS vendors provide are listed here:

- Cloud storage—files/images/blobs

- Database management—NoSQL/**Structured Query Language (SQL)**

- Email services—verification, mailing lists, marketing

- Hosting—web content, **content delivery network (CDN)**

- Push notifications—push messages for Android and Apple

- Social—connecting with social network accounts

- User authentication—social authentication, **single sign-on (SSO)**

- User administration

- Location management

- Endpoint services—**REpresentational State Transfer (REST)** APIs and GraphQL management

- Monitoring

- Analytics, **machine learning (ML)**, and **artificial intelligence (AI)**

- Queues/**publish-subscribe (pub-sub)**

- **Continuous integration/continuous deployment (CI/CD)** and release management

- Tools/libraries/**command-line interfaces (CLIs)**/SDKs

The list varies from vendor to vendor; some specialize in a single or a few related services, while others try to be all-rounders. Firebase and its **Amazon Web Services (AWS)** counterpart, AWS Amplify are examples of vendors that are trying to do everything. Okta provides identity services for all sorts of products and organizations, but they have specialized **identity and access management (IAM)** services that can be used by mobile and web applications. Some of the top BaaS/mBaaS vendors are listed here:

- Firebase

- AWS Amplify

- Okta

- Backendless

- Azure Mobile Apps

- BaasBox

- Baqend

- Kinvey

- Back4App

- MongoDB Stitch

- And many more…

All BaaS or mBaaS services abstract away all resource management from the customer and hence fall into the category of serverless. So, how does BaaS fit into the cloud delivery models we discussed in the first chapter? Does this sound similar to **Platform as a Service (PaaS)** again? That's because BaaS is exhibiting all features of a serverless service. The differentiator here is that, in addition to all the services that PaaS provides, BaaS also provides cloud services, tools, and frameworks that accelerate the building of complete backends for web and mobile applications.

So, are these BaaS services what we can consider as services that will provide serverless backends for FaaS platforms? Only partially. In the context of serverless and FaaS, we will extend the BaaS definition and add all services that can power a serverless workflow into their kitty. This could include a whole bunch of services that are probably not needed to roll out the common web or mobile applications. To get some clarity, we will pick a few services and architectural patterns that can power serverless workflows. Let's begin by looking at messaging systems.

Messaging systems

In a data-driven system, messages contain a unit of information or instructions for a unit of work. Passing these messages from one system to another allows for **inter-process communication (IPC)**, or rather, **inter-service communication**. This allows for the decoupling of various services and forms the basis for event-driven systems, which we discussed in *Chapter 1*. Every message would have a producer creating a message and sending it to the messaging system and a consumer who would pick it up and process it.

In this section, we are going to discuss messaging systems in more detail, how various models of communication differ, and how and when to use them. In messaging systems, there are two architecture patterns—message queuing and pub-sub.

Message queues

A queue is a sequence of items waiting to be processed in the order they arrive. This is similar to the typical **first in, first out (FIFO)** queues in software design. The first item that arrives in the queue is the first one to be processed. The message queue pattern is used in scenarios where **point-to-point (P2P)** communication is required. This means that a single process—aka a producer—creates all messages and places them in a queue, and then one or more processes—aka consumers—picks up each of those messages in the same order and processes it. Messages in such queues can only be consumed once, and once a consumer processes a message, it is deleted from the queue. This delivers a strong guarantee of a message being processed by one consumer, exactly once. This is a suitable model for a set of workers taking instruction from a central controller and executing them only once through a single worker.

Here is a basic diagram of a queue with one producer and consumer:

Figure 2.1 – Queue with no scaling

Here, there is only one producer and one consumer processing messages if this operation has to be scaled up. For this, more consumers can be added. But one message would still be processed by one consumer. Scaling up would send different messages to different consumers but only once, as shown in the following diagram:

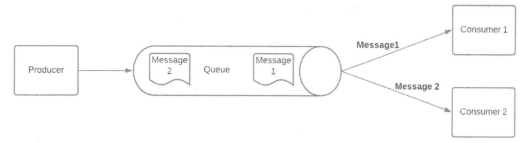

Figure 2.2 – Queue with scaling

The ability to scale in this way means message queues can support high-volume traffic by just adding more consumers. Which consumers consume a particular message will depend on the implementation of that particular message queue. If a particular consumer is unavailable for a while, a message may not be delivered until it is back up, while other consumers would keep processing their messages. This could cause messages to be processed out of order. You need to understand the ordering guarantees provided by the messaging service before implementing your solution.

Pub-sub

The pub-sub pattern differs from message queues primarily in how consumers read the messages. Publishers are the producers of messages, who publish to a *topic*, and consumers subscribe to these topics and consume as and when messages are available. In contrast to how message queues operate, pub-sub systems allow multiple consumers to receive the same message on a topic. Rather than deleting messages, pub-sub systems persist messages to disk or memory and keep them for a while as defined by the retention policy of the system.

Pub-sub systems are a good fit where information has to be passed on to several parties—for example, sending notifications to all traders who have enabled push notifications for a certain stock crossing some value. In most implementations, ordering is preserved but make sure to check this out before onboarding to any pub-sub service. For example, Kafka message brokers provide order guarantee within a partition (a storage abstraction implemented by Kafka) of a topic. A topic could have several partitions, and at any point in time, only one consumer is responsible for consuming from a partition. This provides an order guarantee, not within the topic, but rather, the individual partitions of the topic that are being consumed. You can see an illustration of this in the following diagram:

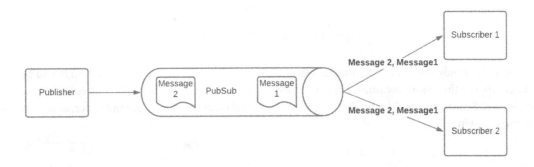

Figure 2.3 – Messaging: pub-sub model

In the coming sections, we will look into some common aspects of messaging systems, such as protocols and client interaction.

Message brokers and clusters

Message brokers are software applications that implement the messaging architectures covered in the previous section. They essentially implement the server component of the messaging system. The features of any message broker include the following:

- Support for one or more commonly used messaging protocols
- Message routing to a destination
- Converting messages to various formats, either for local storage or to send to consumers
- Authentication and authorization
- Aggregating and/or dividing messages for efficiency
- Indexing, compaction

Not all brokers implement all features, but message routing and data marshaling (conversion, storage, and so on) are minimum requirements for brokers.

In a distributed architecture where multiple components rely on asynchronous communication using brokers, having a standalone broker is a bad choice. Apart from being a **single point of failure** (SPOF), it also limits scalability. Hence, almost all broker implementation comes with clustering capabilities. The implementations vary, but most clusters implement some or all the following features:

- **Partitioning of data**—How all messages belonging to a queue or topic will be stored across a set of servers.

- **Replication**—Fault tolerance requires a predetermined number of copies of this data to be stored (called the *replication factor*).

- **Controller/leader**—To manage the replication and enforce consistency across brokers in a cluster, most technologies implement a central controlling unit to orchestrate administrative duties.

- **Inter-broker communication**—Brokers need to talk to each other for data management, replication, leader election, and other mechanisms that a distributed system should implement.

- **Failover protocols**—In the event of a broker going down, temporarily or permanently, traffic and data management have to be handled by another peer.

Some of the most popular broker implementations are listed here:

- Apache Kafka

- Apache ActiveMQ

- Apache RocketMQ

- Apache Qpid

- Apache Pulsar

- RabbitMQ

- NSQ

The following section covers some of the most commonly used messaging protocols.

Messaging protocols

Different broker implementations support different protocols—some of them are custom, while others are open standards. Here are a few of them:

- **Advanced Message Queuing Protocol (AMQP)**

- **Message Queuing Telemetry Transport (MQTT)**—Widely used in IoT

- **Streaming Text Oriented Messaging Protocol (STOMP)**

- **Jakarta Messaging**—Earlier known as **Java Message Service (JMS)**, this is a Java-specific implementation
- **Extensible Messaging and Presence Protocol (XMPP)**—Originally called Jabber, ideal for instant messaging use cases and **peer-to-peer (P2P)** communication

In the next section, we will explore two models of message retrieval methods—push and pull.

Push versus pull

Producers almost always push messages to a broker, but with consumers, various brokers implement either a push or a pull model.

In a push model, the broker pushes a message to the consumer as and when it becomes available. The consumer is passive in this case, and the broker has to actively manage the message delivery. The advantage of a push model is that the client stays up to date as the message is delivered, as soon as it is available to the broker. The disadvantage is that if the consumer is not powerful enough to cope with the rate at which the broker is pushing messages, it will cause backlogs and subsequently lead to the consumer going down. In addition, the broker has to keep track of each consumer and how many messages they have consumed.

A pull model is consumer-centric and allows consumers to poll brokers periodically to get new messages. This is more reasonable, and the consumer can tune their pull logic to suit their needs. It also relieves brokers from the need to manage consumer status and state. A typical problem with a pull model is that consumers can't guess when new messages will arrive. This means that consumers have to come up with a logic to poll the brokers to check for message availability. Finding the right time interval to poll is tricky—make it too short, and you overwhelm the broker; make it longer, and you might miss out on messages that arrived within that span of time.

Kafka and RocketMQ use pull mode whereas ActiveMQ uses push mode.

Guarantees in messaging

Guarantees in messaging are essentially about the reliability of the messaging system to ensure a message sent by a producer to a broker is received by the consumer. A message can be lost in three situations: on the way from the producer to the broker, during storage at the broker, and on the way from the broker to the consumer. Various broker implementations use techniques such as acknowledgments, retries, replication factors, redelivery, delayed delivery, and others to ensure the reliability of messaging during communication as well as persisting in the disk. To quote an example, the Apache Kafka producer provides three choices for delivery guarantees—*at-most-once*, *at-least-once*, and *exactly-once*.

Dead-letter queues

Dead-letter queues (**DLQs**) are catch-all queues specifically designed to hold messages that failed to be processed due to some criteria, as outlined here:

- Queue capacity exceeded.

- Message size exceeded.

- Destination queue or topic doesn't exist.

- The message has expired due to **time-to-live** (**TTL**) expiry.

- The consumer fails to process the message.

Different brokers implement DLQs with different rules. DLQs allow application owners to set monitoring on the DLQ to find issues with the system or the processing logic. This improves the reliability of the system by resolving systemic issues as well as design faults.

Cloud messaging

Running a message broker cluster is an operation-intensive task. You may run into maintenance issues not just because of the systems or broker software but also due to the usage patterns of producers and consumers. Since messaging systems are stateful systems, horizontal scaling is not as easy as web application scaling. In short, while the advantages of asynchronous architecture are manifold, running a messaging system comes with its own engineering cost.

Cloud-hosted messaging systems solve the problem of maintaining and scaling your messaging systems. All of the infrastructure aspects of messaging are abstracted out, and only an interface to talk to the messaging system is exposed. All administrative tasks that are available to the customer will be exposed via cloud consoles or APIs. The interface and the configurations available for the customer to tune depend on the vendor and the specific offering. Most cloud vendors provide more than one type of messaging system, some of which offer support to traditional messaging protocols such as AMQP, **Java Messaging Service** (**JMS**), and STOMP while also providing homegrown solutions with their APIs and SDKs. There are also vendors specializing in providing **Messaging as a Service** (**MaaS**). Vendors who provide IoT services offer messaging systems that support the MQTT protocol.

Some cloud messaging services provided by major vendors are listed here:

- **AWS**—Amazon MQ, **Amazon Simple Queue Service** (**Amazon SQS**), **Amazon Simple Notification Service** (**Amazon SNS**), and Amazon IoT Message Broker

- **Azure**—Azure Service Bus, Azure Event Grid, Azure Event Hubs, and Azure Queue Storage

- **Google Cloud Platform** (**GCP**)—Pub/Sub and Pub/Sub Lite

Cloud messaging platforms are essential for almost all complex serverless applications. Responding to and processing events requires messaging as a central component, but compare and contrast different messaging offerings from vendors and decide which one fits your business use case before starting to use it, as each platform has a different set of features.

Object stores

Object-based storage (aka object storage) is a data storage architecture where data is managed as objects. Traditional filesystems use block storage where data is stored as files and organized in folders. Under the hood, the files are linked lists of data blocks stored on top of a block filesystem, which in turn is built on top of disks.

Object storage is ideal for storing large amounts of unstructured data. *Unstructured* doesn't necessarily mean that the data in this storage has no structure. Rather, it doesn't conform to the data format that can be fit into normal structured data storage systems such as **relational database management systems (RDBMS)** or NoSQL. The internet is full of such unstructured data, with images and videos topping the list. Web pages and text content, including emails, come next. While most of this content has its own format, it is not easy to store, search, or retrieve using traditional data storage methods.

An explosion in the volume of unstructured data produced across the internet has led to challenges in storing and archiving data effectively. Object stores are also being used for the backup and archival of structured data, such as RDBMS backups. Traditional filesystems are reaching their limits, and object storage has emerged as the viable solution. Let's examine how object storage works.

Design and features

The fundamental principle of any object storage is how files stored in the underlying storage system are abstracted out. Users, or even administrators, have limited-to-no control over how data is stored. Let's look at two basic building blocks of object storage —namespaces and objects.

Namespaces

This is an abstract concept and is termed variously in different implementations and cloud platforms. The fundamental idea is that objects have to be organized based on the use case, and there needs to be some sort of a container on top of objects. Ideally, object stores have a flat layout and don't entertain hierarchies such as a filesystem directory structure. The namespace/container, or whatever you want to call it, is the only top-level container for all objects.

AWS Simple Storage Service (S3) is one of the earliest cloud object storage providers. Its namespace is called a bucket and is a globally unique name. In the open source world, **OpenStack Swift** is an object store with a *container* as the namespace for objects. But unlike S3, containers are not globally unique and are organized under the **account identifier** (ID), which is a unique organizational construct within OpenStack. Irrespective of the implementation, objects always have a flat hierarchy under a namespace.

Objects

Objects essentially store data content. They are mostly files that we deal with in internet-scale businesses—video, audio, images, archive files, logs, and so on. Each object is identified with a **unique ID (UID)** within the namespace. This UID enables the user to get an object without knowing anything more about the data layout underneath. This is also helpful for providers to abstract out storage implementation or change it without disturbing the customer. This allows for expanding the infrastructure and provides virtually unlimited storage, without the customer ever having to worry about what happens underneath.

Even if an object store has a flat structure under namespaces, in many implementations, the object ID can contain forward slashes, which can help to organize objects in a file/folder structure. The namespace and object together form the most vital feature of object storage—a globally unique **Uniform Resource Locator (URL)** for each object. This simplifies designing and developing the business logic around objects, and the application designers no longer have to worry about the underlying storage semantics.

An object is usually made available to clients via a REST API, making it simple to access— as well as administratively manage—objects. The usual **HyperText Transfer Protocol (HTTP)** methods of PUT for addition, DELETE for removal, and GET for retrieving work seamlessly. Authentication and authorization are also easier to be used with REST. Generally, most vendors—either cloud or open source—provide their own software development kit in different languages to interact with the REST endpoint, making it much easier to develop code.

Both objects in object storage and files in block storage have metadata associated with them. Metadata is information that provides more context about the data that is in storage. For a regular file in a filesystem, this could include ownership information, size, permissions, filesystem, and block-level information. In a given filesystem, there will be a limited set of metadata that will be supported, and for the most part, users can't modify any of this metadata. This is where the object store has a significant difference. While object storage provides a basic set of metadata by default, it also empowers users to add their own metadata. This metadata is usually in a key-value format and helps in providing additional context about an object. It can be modified, extracted for further processing, and be used for categorizing data. Running analytics on your video or image files is a good use case, assuming the applications storing these files as objects have additional context information that helps analytics.

Benefits of object storage

Object storage provides a cloud-native storage option that provides more flexibility in storing and retrieving content as compared to traditional block-level storage. Let's look at some advantages of object storage here:

- Practically infinite storage—**exabyte (EB)**-scale
- Accessible from anywhere using the HTTP protocol
- Scale transparently to accommodate any scale of requests

- Multiple redundancy options and replication policies

- Better data analytics based on metadata

- Store very small to very large objects

- Data secured in transit and storage

- Cheaper than block-storage solutions

- Faster data retrieval

- Programmatic data management via REST APIs

Object stores are best suited for data that is often read but rarely modified. They are not ideal for real-time systems since most object storage doesn't provide the guarantee that a read immediately after a write would return the most recent data.

Let's look at some of the objects stores available, both in the cloud and open source.

Open source object stores:

- Red Hat Ceph

- OpenStack Swift

- MinIO

- Riak S2

- OpenIO

Cloud object stores:

- AWS S3

- Azure Blob Storage

- GCP Cloud Storage

Since AWS S3 was one of the earliest cloud object stores and became widely popular, most open source and cloud object stores support cross-compatibility with S3 protocols.

Object storage is used in all cloud and non-cloud workflows nowadays. With the advantages described in this section, it becomes the perfect candidate for the file backend of serverless applications. In the next section, we will explore the concept of workflows and workflow automation in the context of cloud and serverless platforms.

Workflow automation

Business and **information technology (IT)** processes can often get very complicated. They involve multiple tasks in multiple stages, each task taking input from previous tasks and external systems and making decisions based on the output of various steps. These are boring, lengthy, and repetitive tasks that can and should be automated.

These multi-step processes—also called workflows—are a core part of any business. Consider a **financial technology (FinTech)** application approving a loan application. These are a bunch of processes the bank has to follow. Let's cherry-pick some of those processes to give an idea of how business workflows work. Have a look at the following diagram:

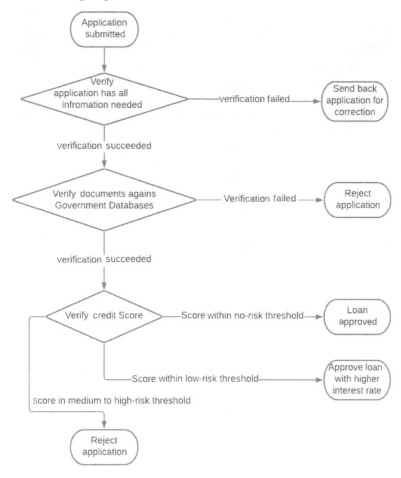

Figure 2.4 – Loan approval workflow

So, this becomes a business process with success or failure in each stage defining what to do next. As long as government ID databases and credit scores are available over an API, this process can be fully automated. In the case of the step where the application doesn't have the necessary information and is sent back, either the loan officer and/or the applicant will get a notification to initiate the process again. While reinitiating is a manual process, the scope of the approval process can be limited to sending the application for correction, and subsequently notifying the customer and loan officer through digital channels, post which the approval process can be closed.

Traditionally, the way businesses handled this automation was either by building a full in-house application or—more often—using tools that are built to automate workflows. These tools are, in general, called business process management software. Fundamentally these tools provide a way to visually organize and define a workflow and then run the automation using low-code or no-code options, depending on whether the workflow steps are standard or custom. There is also a standard called **Business Process Model and Notation (BPMN)** that defines standard notations and formats for visually organizing a workflow. Without going into the details of the standard, this could be considered a more sophisticated version of a flowchart to define business processes. There is software that implements BPMN—one such popular open source tool is **Camunda**. Written in Java, it allows for the production of programmatically defined workflows and visualization of the process via a **user interface (UI)**.

While business process automation is vital for the business, there are also other critical workflows such as IT department automation processes. The same logic and software can be used to automate IT workflows as well. This idea has given rise to several open source and cloud-based workflow automation frameworks. While open source tools automate each step with code blocks or modules written in the same language as the one in which the workflow management software is written, cloud workflow automation frameworks turn to FaaS for this functionality. They also allow event triggers and inputs from various BaaS systems. Some popular open source workflow management software are listed here:

- Camunda
- Airflow
- Luigi

There are specialized workflow managers that are tuned for specific IT needs. For example, Oozie and Azkaban are workflow managers used in the Hadoop ecosystem to orchestrate data pipelines. The cloud version of workflow management achieves similar results, but without the hassle of having to manage the infrastructure and worry about the inner workings.

Cloud workflows

Cloud workflow systems allow you to run serverless workflow orchestration in the respective platforms. These workflow services can be used for automating cloud service interactions, business processes, and microservices workflows. They essentially allow you to execute rule-based logic—which forms the steps of the workflow—to solve IT and business problems. The advantage of cloud workflow systems is that most cloud services will be integrated into the workflow service, making it easy to respond to cloud events or take input and post output to cloud data systems. There is no need to do capacity planning for the workflow execution.

In most cloud workflow services, the business logic to be executed in each step is run using functions written in their own FaaS platform. This also allows developers to pick any language that the FaaS platform supports—or a combination of them—for each step, adding flexibility and easing the reuse of code. Since the automation orchestration is taken care of by the workflow engine, there is very limited code that customers need to write in order to connect the steps.

There are a lot of use cases that can be onboarded to cloud workflow services, many of which depend on the level of integrations offered by particular cloud platforms, but a few common use cases are outlined here:

- **Extract, transform, and load** (ETL) pipelines
- IT infrastructure automation
- Compliance management
- Application tracking
- Automated alert remediation
- Responding to security events
- Business processing and approvals
- Data pipeline management

Workflow management services provided by major cloud vendors are listed here:

- AWS Step Functions
- Azure Logic Apps
- Google Cloud Workflows

The relevance of a workflow management service is that it can be integrated with most microservices architectures. Most business use cases have to follow some part of the process, and offloading part of the application logic to serverless workflow management systems makes software development easier and agile. A lot of backend code can be replaced or augmented by integrating cloud workflow engines.

In the coming section, let's explore an important data platform/architecture concept that is vital for storing data in the cloud, and serverless in particular: NoSQL.

NoSQL platforms

NoSQL refers to non-SQL, which is an alternate method for modeling, storing, and retrieving structured data. The traditional method is storing data structured in an RDBMS model. The method to query a relational database is called SQL, and hence the name NoSQL. While a detailed comparison of SQL and NoSQL is outside the scope of this book, we will have to look at some aspects of NoSQL platforms before discussing cloud NoSQL platforms.

The power of RDBMS databases came from the relationships defined between these databases and the ability to enforce **atomicity, consistency, isolation, and durability (ACID)** properties. At the same time, the relationships made the RDBMS system enforce structures in the form of tables and relationships between the database for all their datasets. This table-and-relationship definition is called a *schema*, which fundamentally defines an RDBMS database.

With the advent of internet-based business models, the demand increased for faster and flexible databases that could handle large datasets with more unstructured data. This is where NoSQL databases came into the picture. With NoSQL, there is no rigid schema, and how the data is interpreted largely depends on the clients accessing it. It could have some structure or loosely defined schema but would still have the ability to accommodate an evolving schema for new data. Consider the case of a social media profile. Not every profile will provide all information, which would cause an imbalance in the structure of each profile. If we were to store this data in RDBMS, it would leave a lot of columns empty across all records, causing inefficient storage and retrieval. If NoSQL were used in its place, there wouldn't be any problem storing the data since there would be no enforcing of structure. It would also come in handy when the social media profile had to accommodate new data items in their profiles since there is no need to go back and modify the database schema or prefill the column for existing records.

This is just one specific use case of NoSQL; some other use cases include personalization, content management, catalogs, e-commerce data, user preferences, personalization, digital communication, and IoT. There are different types of NoSQL databases suitable for different purposes. Let's look at the most important ones here:

- Document store:

  - Store data as documents.

  - Each document has a unique key associated with ease of retrieval.

  - Semi-structured data.

  - Data is encoded in formats such as **JavaScript Object Notation (JSON)**, **Binary JSON (BSON)**, **Extensible Markup Language (XML)**, and **YAML Ain't Markup Language (YAML)**.

  - Use case—social profiles, blogs, and user-generated content.

  - Example—MongoDB.

- Search:

  - Specifically built for indexing and searching data.

  - Stores semi-structured document blobs.

  - Use case—searching logs and metrics.

  - Example—Elasticsearch.

- Key-value:

  - Each data item contains a key and corresponding value.

  - Keys will be unique within the collection of data.

  - Many implementations store data in memory for faster access.

  - Some persist data to the disk.

  - Use case—transient user data, database values, caching read-only data.

  - Examples—Couchbase, Redis, Memcached.

- Graph

  - The relationship between data items can be represented as a graph.

  - A large number of relationships, but finite.

  - Contains edges, nodes, and properties.

  - Edges are relations, while nodes are objects.

  - Use cases—social relationships, IT infrastructure relations, maps, network topologies.

  - Examples—Neo4j, FlockDB, InfiniteGraph.

- Wide-column

  - Use tables, rows, and columns—like an RDBMS.

  - Names and formats of columns can vary from row to row.

  - Related columns grouped into column families.

  - Use cases—sensor logs, time-series data, real-time analysis.

  - Examples—Cassandra, Bigtable.

There are other lesser-used NoSQL types, but they are not very relevant to our context. One type of database to call out is multi-model databases. These are DBMS that offer multiple data models against a single core database backend. Examples of some popular multi-model databases and their supported data models are provided here:

- **ArangoDB**—Document, graph, key-value
- **OrientDB**—Document, graph, key-value, reactive, SQL
- **Redis**—Key-value, document, time-series

So, why is the NoSQL model so popular, apart from the support for a wide variety of data models? Let's look at some advantages of NoSQL platforms here:

- **Flexibility**—Flexible schemas, ideal for semi-structured and unstructured data.
- **Versatility**—A large number of data models to choose from.
- **Scalability**—Distributed systems with scale-out architecture.
- **High performance**—The data models and access patterns enable high performance compared to RDBMS.

While it is comparatively easier to manage NoSQL platforms and their scaling, it still takes a considerable effort to manage them at scale. Depending on the architecture of the NoSQL platform you are using, you will have to take care of read and write quorums, setting up separate endpoints for read and write, optimizing replication, and so on. A cloud NoSQL platform abstracts out all these efforts for the customer, as you will see in the next section.

Cloud NoSQL platforms

NoSQL databases are offered by all cloud vendors. Some of the solutions offered are natively developed by the cloud vendors, but they also provide some of the open source solutions as managed services. With managed services, most scalability and management concerns are abstracted out for the customer, but there could still be some limitations due to the nature of the database product and its inherent scaling models.

With cloud NoSQL platforms, all the advantages any cloud product has will also be available for NoSQL platforms. These include managed scalability, security patching, access restrictions, automated upgrades, and migration services. In most platforms or products, the customer will get to choose certain capacity-related parameters such as expected reads/writes or storage capacity so that the vendor can auto-provision database instances tuned to the customer's needs. Beyond that, everything is managed by the vendor. In addition to this, some cloud vendors also partner up with external vendors specializing in NoSQL and provide a solution with scaling and other aspects managed in collaboration with the external vendor.

Let's look at the NoSQL products offered by major cloud vendors and the type of database they support. They are presented in the following table:

	AWS	Azure	GCP
Document	DocumentDB	Azure Cosmos DB	Firestore Firebase Realtime Database
Key-value	DynamoDB	Azure Cache for Redis Azure Cosmos DB	Cloud Bigtable
Wide-column	Keyspaces (Cassandra)	Azure Cosmos DB	Cloud Bigtable
Graph	Amazon Neptune	Azure Cosmos DB	Managed Neo4j
Search	Managed Elasticsearch	Azure Cognitive Search	Managed Elasticsearch
In-memory	ElastiCache (support for Redis and Memcached) Amazon MemoryDB for Redis	Azure Cache for Redis	Memorystore (managed Memcached and Redis)

Table 2.2 – Cloud NoSQL offerings

As we saw in this section, there are different NoSQL platforms for every need. While some try to cover all use cases, some try to do one thing and be the best at it. Do evaluate your current and future business needs before choosing a platform.

In the next section, we will learn about edge computing and how it helps in moving computing and data closer to the user.

Edge computing

Edge computing is a computing and networking paradigm whereby computing and data are brought as close as possible to the origin of the data and/or the users of the data. The term was originally coined in the context of CDNs. They are placed outside data centers, but closer to the **internet service providers (ISPs)** of the end user so that companies can cache and deliver static content such as images, videos, and scripts to the customers faster. To understand the concept of edge, let's look at the following network diagram of a typical internet service:

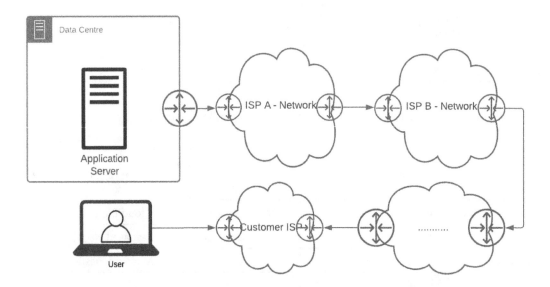

Figure 2.5 – Network diagram of a regular web application

As you can see, there are multiple public networks between a server running an internet service web application and its user. These public networks are owned by various internet providers who all communicate with each other using **Border Gateway Protocol** (**BGP**) and transfer data. There will be multiple paths via different combinations of provider networks. Out of this, which route the traffic between the web application and the customer takes depends on a whole lot of scenarios, including the network routing contracts between ISPs, responding to sudden changes in network topologies due to overload or cable cuts, and so on. However, there are safeguards against most of these issues, and the network can be largely reliable here.

The last network in this chain is the one that connects the customer with their ISP. Part of this ISP network that reaches—mostly via cables—to the router at the end user's home or office is called the *last mile*. It is the final leg of the communication to the customer. This is where the network is the most unpredictable and unreliable, for various reasons, as outlined here:

- The last-mile connectivity has to go through public locations where there is a good chance of physical damage, external signal, and noises.

- In addition, electric power that powers the network devices on either side of the last-mile connection could be unreliable and without backup.

- The network device at the customer location—called **customer-premises equipment** (**CPE**)— could be unreliable due to low hardware configuration, performance issues, and so on.

Given that applications that provide services through the internet don't have any control over the last-mile network, the next best thing they can do is ensure more stability and speed in delivering data and computing services closer to the last mile. This is where edge computing comes into the picture.

The term *edge* can be a bit hazy since the network between the application's server and the end-user system has many components. A reasonable definition of *edge* would be any device in the network path between the application server and the customer or end-user systems on which the application infrastructure (of the business that runs the application server) has control to transform data and/or run compute. This allows the application server to offload a lot of data processing and compute to edge devices, which otherwise would have to be handled by the application server itself.

So, how is edge computing relevant to serverless? Unlike the backend services we discussed earlier in this chapter, edge devices are outside the common cloud infrastructure. Not all edge computing use cases can be handled by serverless, but certain common use cases help speed up the business logic of application servers. These use cases are mostly implemented within the edge infrastructure of cloud providers— to understand this infrastructure, we need to learn a little bit about **point-of-presence** (**PoP**) and edge location concepts in network architecture.

PoP

The core infrastructure of any business runs in data centers, be it directly using bare-metal servers or via clouds. These are huge infrastructures with hundreds of thousands of servers and network equipment. Due to the sheer size of the data centers, as well as the need for the application infrastructure to have all its components in cross proximity, businesses run their applications in a handful of data centers in selected geographical locations. They have to make this choice even if their customers are all across the globe, due to cost and technological constraints, but this has the side effect of application servers being far away from their customers (in terms of networks to pass through) and increases the time to respond to customer requests. To improve the stability and performance of communication between customers and application servers, the network infrastructure of the business has to be placed as close as possible to the customer.

To take the network infrastructure closer to the customer, edge locations have to be identified across the globe. These locations are chosen based on the distribution of customers in a particular region, as well as the budget for such a project. These locations are mostly **internet exchange points** (**IXPs**) or other peering locations of top-tier internet providers of that region. Without going too much into the networking terminologies, these are locations where various ISPs connect their networks to form the internet. The advantage of these locations is that they will have huge bandwidth and direct network connectivity with most regional and global ISPs, which reduces the latency from these edge locations to last-mile networks.

Once edge locations are identified, a dedicated internet line with high bandwidth is established between the data centers and the edge locations, thus adding stability and reducing latency. Now, any traffic that will come from the customer and passes through the edge location will reach application servers in data centers with more stability and speed. To have customers come through the edge locations closer to them, different routing and **Domain Name System (DNS)** techniques are used, which are beyond the scope of this book. Assuming the users are directed to the nearest edge location, companies need to establish their edge servers and network gear to form a miniature data center. This mini-data center is called a PoP. The customer traffic is directed to the servers running in PoPs, which will do initial processing, establish **Transmission Control Protocol (TCP)** connections, and **Secure Sockets Layer (SSL)** termination, helping reduce network delays. These servers often perform the functions of layer 4 load balancers and layer 7 network proxies. After the initial processing, the customer request is forwarded to the application servers in the data centers via a fast, dedicated network link between the PoP and data center. The following diagram will help in putting the concept together:

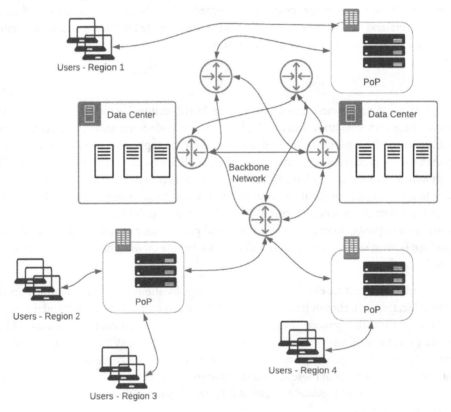

Figure 2.6 – PoPs and data centers

Now that we have a foundational idea of how PoPs work, let's look at one of the earliest use cases of edge infrastructure—CDNs.

CDNs

A CDN is a geographically distributed set of proxy and cache servers that accelerates the delivery of web content. A large amount of data served by applications these days is static—videos, images, scripts, and so on. These contents are usually referred to as static assets and don't change very often. The same file will be shared between a large number of users. Consider the case of a YouTube video or Facebook profile photo. Serving these static assets from application servers in data centers (which are limited in number) and transporting them across long distances to deliver to the customer each time wastes a lot of network and computing resources. Instead, companies use the concept of edge and bring static assets closer to the customers. To do this, CDN providers establish PoPs in locations across the globe, set up servers that can cache the static assets in their storage, and act as a proxy to application servers serving the static content.

Consider a typical scenario of a user from Bangalore, India requesting a video. The application servers would have this video stored within their data center infrastructure. When a user requests this video for the very first time from Bangalore, the request goes through the PoP closer to Bangalore, and it looks for the video in the caching servers within the edge location. Since nobody had requested this video before, it won't be in the cache, and hence the request will be forwarded to the application servers in the DC. Application servers (called origin servers in a CDN—servers where the content originated from) require the video to be transmitted to the user via the PoP, and in the process, edge caching servers will also cache the video in their local storage. The next time a request for this video comes into the PoP, it will be served from the local cache rather than going all the way to the DC and back. This greatly speeds up the time to deliver static assets. Consider this video is requested by thousands of users from Bangalore or locations closer to Bangalore. The benefits add up, and that's the value provided by the CDN.

The following diagram shows how CDNs can help speed up serving content delivery:

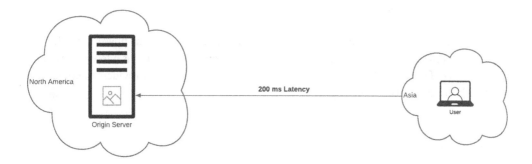

Figure 2.7 – Requesting an asset from an origin without a CDN

As you can see, requesting an asset from a web server directly by the client is straightforward and involves no complexities. The only downside is that it incurs latency and resource utilization for fetching the asset from the server, which might be halfway across the globe. In the next diagram, you can see how introducing CDN locations closer to the customer helps in reducing the latency described previously:

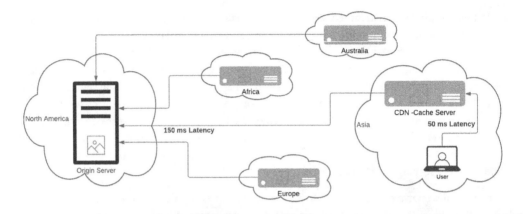

Figure 2.8 – Requesting an asset from an origin with the image cached in a CDN

Not all businesses can afford to build PoPs and CDNs—only a handful of internet-scale companies build those for themselves. Everyone else makes use of managed services provided by dedicated edge and CDN providers. In the case of the cloud, all cloud vendors maintain their own edge infrastructure—they build a large number of PoPs across the world, allowing them to terminate the customer traffic closer to their home locations. They also build a CDN on top of it, which allows static assets hosted in cloud DCs (usually called zones and regions) to be served from these locations. Contents stored in cloud object stores can be easily served up through the CDN. But serving content and terminating network connections are not the only functions cloud edge locations serve—this is where edge serverless services are implemented.

Serverless at the edge

Having computing and storage placed at edge locations provides vendors with the power to implement edge computing—processing data and running compute at the edge. Edge locations intercept customer traffic, which most of the time is HTTP and other web protocols. This allows providers to implement some of the following services at the edge:

- Authentication
- Request validation
- Request manipulation and rerouting
- Modifying cache behavior

- Security checks

- Implementing **web application firewalls (WAFs)**

Many of these services work in tandem with backend services in the cloud, but what supercharges customer request processing at the edge is the ability to run serverless functions as part of the request processing workflow. This usually is an extension of the FaaS platform provider by the cloud, but with some restrictions to efficiently run it at the edge. As of this writing, out of the top three cloud vendors, only AWS provides serverless functions at the edge. There are two flavors of this service—Lambda@Edge and CloudFront Functions. Lambda@Edge is more powerful, with support for both Python and JavaScript and access to the cloud backend services such as DynamoDB for more complex workflows. CloudFront Functions, which was launched in 2021, involves more restricted, ultra-fast functions supporting JavaScript. It is used for manipulating requests and responses with short-lived and fast functions that can only operate on request and response and no other data sources.

Azure and GCP will hopefully follow AWS in introducing functions at the edge, given their huge potential. Meanwhile, there are other players in the edge functions arena, the two top vendors being Cloudflare and Akamai. Cloudflare workers and Akamai edge workers are FaaS platforms deployed at their edge locations and allow programming in JavaScript. Being CDN and network-focused companies, these vendors have powerful edge infrastructure to support the FaaS platform. In addition, both of them also provide high performant key-value stores that can be used as a backend for edge serverless functions. We will look into Lambda@Edge and Cloudflare workers in more detail in the upcoming chapters.

IoT is a special use case and an increasingly popular one in today's world. In the next section, we look into IoT clouds.

IoT clouds

IoT is a network of a variety of physical devices—things—that are connected to the internet. These devices could be smart-home appliances, such as fridges and washing machines, manufacturing devices, heat and motion sensors, smart cars, and self-driving cars . Most of these devices will generate sensor data about their environment or the equipment in which it is embedded, which can be sent to a central location for processing. In some cases, these devices also take instructions from a central system. The sensor data gathered could act as the input for decision-making done by this central system.

IoT devices generate humongous amounts of data, and processing this data on the devices or even on their premises may not be feasible. This is where IoT clouds come into the picture. IoT clouds are hosted solutions that enable a large number of IoT devices to be managed centrally, along with providing support for data transport and processing. These clouds could be part of general cloud platforms from vendors such as AWS/Azure/GCP or purpose-built dedicated IoT clouds. Some common features offered by them are connectivity and network management, device management, data processing and analytics, monitoring, visualizations, and so on.

Most IoT clouds networks would have an IoT gateway to do some local preprocessing or aggregation before data is sent to the cloud, but even without an IoT gateway, it is easy to send data to IoT clouds. Most clouds provide a cloud messaging middleware that supports protocols such as MQTT that can accept data pushed from IoT devices and then send it to other cloud platforms for processing.

So, how is IoT relevant for serverless? IoT data processing requirements oftentimes align with serverless features. IoT devices have varied behaviors—some sending a barrage of data all the time, some waking up at defined intervals to collect and send data, and so on. Most of the time, the data sent and processed is dependent on seasonal events and will have peak and low times. Consider the example of traffic cameras during peak office hours versus noon or manufacturing devices during a plant's shift hours versus off-hours. Processing such events in the cloud requires rapidly scaling computing services. It should also be able to scale down completely when IoT devices are hibernating or minimally operating. This type of architecture works best with serverless and FaaS models. IoT devices can push data into the cloud via IoT middleware messaging services, which in turn will be pushed to various data systems. Functions can then be used to respond to these events and process the data accordingly.

Covering the IoT use case is outside the scope of this book as it requires a lot more focus on how various IoT devices and protocols work and learning how to integrate them to various cloud IoT platforms. This would be a detour from our plan to discuss various cloud and on-premises serverless platforms and hence can't be undertaken, but this section should give you a minimal introduction to what the IoT ecosystem is and the relevance of serverless in managing IoT workloads in the cloud.

In the next section, we will look into stream processing architecture patterns and what is their relevance in serverless.

Stream processing

We discussed event-driven architecture in the first chapter. Stream processing is a specific flavor of event-driven architecture. In general, there are three styles of event processing, as outlined here:

- **Simple event processing (SEP)**—A simple event is an occurrence of something significant that can trigger an action. These events are independent and are ideal for coordinating the real-time flow of work. Consider the case where an event is triggered when a file is uploaded to an object store and that event triggers adding some customer headers to an object. This event requires immediate action and is independent of any event that occurred before or after that.

- **Complex event processing (CEP)**—In complex event processing, multiple events have to occur before an action can be taken. Filtering, selecting, and aggregating multiple events might be required before correlating and concluding that an event of business interest has occurred. The events involved may not be similar and might occur at different points in time. Consider the example of anomaly detection in web traffic coming into a particular web application endpoint. Consider some common anomalies—an abnormally large number of requests coming from a single **Internet Protocol (IP)** address, or an attempt to use the same cookie from a large number of locations. Before flagging these as anomalies, we might also have to cross-check if this is organic or seasonal traffic across all web endpoints or whether a new release has been rolled out or data center failover has occurred. These patterns can't be identified from independent events and instead require combining and analyzing a large number of requests over a defined period.

- **Event Stream Processing** (ESP)—Streams are continuous flows of events in a defined order, generated by one or more systems. Unlike batch processing, where a large dataset is periodically processed, stream processing acts on a steady flow of events. Batch processing is bounded, whereas stream processing is unbounded. SEP enables responding to business events in real time and helps in information flow throughout the enterprise. The usual use cases are real-time analytics, anomaly detection and response, IoT telemetry processing, application logs, website activities, and so on.

Stream event processing is crucial in today's business applications, as almost all applications produce large quantities of flowing data. Consider the case of tracking clicks on a web application. There will be different consumers for such a stream of data, such as advertising platforms, user-engagement analysis services, or fraud-detection tools. Each of these applications becomes consumers of the steady stream of click events—the stream in itself doesn't bother about the consumers and which point in the stream they are reading from. This is for the application to figure out, and that makes it easy for streaming platforms to deliver millions of events at scale. The cloud streams will scale automatically for changing needs and don't require any sort of provisioning, making them an ideal backend for serverless stream processing. The consumers of streams also have to be scalable as well; while you can build microservices to process events and configure scaling, it is much easier when FaaS is integrated as a consumer of the platform.

Cloud streaming platforms provide not just a way to stream events, but rather build an application ecosystem that supports common use cases out of the box. For example, AWS provides Kinesis as a streaming platform. It supports the collecting, processing, and analysis of streams in real time. Kinesis Data Streams can collect large streams of data with low latency and persist it for long periods, up to 1 year. Similarly, Kinesis Data Firehose can accept events from various sources and pump them to various storage solutions such as data lakes, search, and database services. Kinesis Data Analytics helps you process data from various streaming services, including Kinesis Data Streams and Managed Kafka Service, and push the processed data to analytics tools. For building your application on top of the Kinesis data streaming platform, Lambda can be used, resulting in a complete serverless streaming solution.

As you can see, cloud streaming platforms are much more than simple streaming systems and add value by providing support services that help in integrating and processing events, resulting in completely cloud-based streaming and analytics services with minimal or no coding required. Similar to Kinesis, Azure has Azure Event Hubs and Azure Stream Analytics. GCP has Dataflow and streaming analytics. All these platforms provide ways to integrate with analytics services and build pipelines in a serverless manner. While these options are great cloud solutions, if someone wants to get started with open source alternatives, there are some great choices. Apache Storm, Apache Samza, and Apache Flink are on top of the list. Kafka Streams and Apache Pulsar are also good choices.

Future of serverless

One trend that we see in terms of FaaS is support for more and more programming languages. There are more runtime supports than ever, and the trend will continue. Many of these serverless trends—as in other cloud services—are set by AWS, so it would be fair to assume some of the recent developments in AWS Lambda will be followed by other vendors as well.

Trends in serverless

There are new use cases and technologies that are shifting to a serverless model. Here is a list of trends in serverless—while this is not a comprehensive list, it is a speculation based on how various vendors are moving in the serverless space:

- **Ability to use custom containers**—FaaS platforms in general offer packaging of functions and their dependencies, while the underlying runtime and container image remain the same. Customers have no ability to influence the container and runtime. This has changed for Lambda and will change for other platforms as well.

- **Memory and central processing unit (CPU) time limits**—Lambda and other FaaS platforms usually have limits on how much memory or how much time a function can use in one run. Initially, FaaS was considered a short-lived compute function, which is still the case for most workloads. But increasing these limits allows users to build long-running microservices within Lambda rather than deploying on their own.

- **Observability**—All cloud platforms provide their own services for monitoring and metrics for FaaS, but this is an area where a lot more improvement can happen. Either the cloud vendors themselves or third parties will start adding tooling that will provide better visibility into serverless workloads.

- Security is always a central theme in the cloud, as a small misconfiguration will enable threat actors to take over your cloud resources. FaaS also suffers from the same problem—there will be more customizable security policy support for FaaS and BaaS in the future.

- A need for more customized and powerful serverless platforms will lead to running self-hosted and open source serverless platforms. Kubernetes will be a catalyst in this move.

- Stateful systems will increase adopting FaaS to process part of their workload, as well as adopting more BaaS platforms for better scalability.

- Serverless workflow and business process automation will become an integral part of serverless computing.

- More serverless vendors will come up, other than the traditional cloud vendors.

- The need for multi-cloud and hybrid serverless models will increase.

- Further interoperability and integration between BaaS services will occur to provide low-code and no-code platforms that are self-sustainable.

- There will be an investment in optimizing FaaS models with newer lightweight virtualization techniques, such as AWS Firecracker.

- Services will move toward more event-centric integrations, and serverless and FaaS will play a central role in this.

- The cost and performance of each piece of code will become a first-class citizen in serverless application design.

- **Serverless RDBMS**—Currently, AWS Aurora is the only vendor to offer a serverless RDBMS, but more vendors will get into this domain.

Serverless has already established itself as a powerful cloud computing model. More and more vendors are getting into serverless platforms and FaaS, whereas the current players are adding more integrations and interoperability between serverless services. Serverless as a whole is a combination of FaaS—which is stateless compute—and BaaS, which is stateful. This helps in building scalable workflows around various data stores, accelerating application development and faster **time to market** (**TTM**). The future will see more powerful and integrated serverless solutions.

Summary

In this chapter, we covered BaaS and explored the most important backends and architectural patterns that can be used in serverless workflows. Using object stores and NoSQL platforms for data management in serverless empowers it to build real applications using FaaS. With an ever-increasing flow of business events, stream processing and message queues will play a vital role in solving business problems the serverless way. Edge computing and IoT take data and computing closer to customers, paving the way for increased automation and data distribution while improving the customer experience. Workflows are the lifeblood of all businesses, and workflow automation and cloud workflows take it to the next level.

Chapter 1 and this chapter have so far provided enough foundational knowledge on serverless. It has mostly comprised theoretical and 1,000-feet overviews to set the stage for the technologies that we will cover going forward in this book. In the next chapter, we will be looking into AWS and its serverless stack. Toward the end of the chapter, we will also implement a project with chosen serverless technologies of AWS.

Part 2 – Platforms and Solutions in Action

This part provides a deeper look into the serverless platforms and services offered by various cloud vendors. We will discuss specific technologies within each platform. Then, we will move on to design and implement a serverless project for each platform using a few of the technologies in that platform. The chapters in this section are independent and can be read according to your preference.

This part has the following chapters:

- *Chapter 3, Serverless Solutions in AWS*
- *Chapter 4, Serverless Solutions in Azure*
- *Chapter 5, Serverless Solutions in GCP*
- *Chapter 6, Serverless Cloudflare*
- *Chapter 7, Kubernetes, Knative, and OpenFaaS*
- *Chapter 8, Self-Hosted FaaS with Apache OpenWhisk*

3
Serverless Solutions in AWS

Amazon Web Services (AWS) is the oldest and biggest vendor in the cloud landscape. What humbly began as a service offering three core infrastructure services over the internet – **Simple Storage Service (S3)**, **Simple Queue Service (SQS)**, and **Elastic Compute Cloud (EC2)** – has expanded to provide over 200 services across the globe. With the advantage of being the first entrant into the field, as well as the size of its infrastructure, AWS often sets the trend in the industry, introducing new technologies as well as cloudifying existing on-premise services. The case of serverless is no different. When AWS started offering Lambda, its FaaS platform, it created a trend and was quickly adopted by a large number of customers for a wide variety of use cases. While Lambda in itself was launched not so long ago, a lot of the backend services that fall into the BaaS categories we covered in *Chapter 2* have already been in production for a long time, such as S3, DynamoDB, and SQS. After the launch of AWS Lambda, these battle-tested services were integrated with Lambda in various ways to create more elegant solutions with less coding.

In this chapter, we will cover some of the core AWS serverless services and then build a product based on these services. We will cover all the essentials required for you to get a fair idea of each service, but coverage of all features and tunables of each service is beyond the scope of this book. Having said that, we will provide you with enough knowledge that you will completely understand the project we will implement so that you can do research for further understanding.

The main topics we will cover are as follows:

- Fundamentals of cloud formation
- Lambda
- API Gateway
- S3 – object store
- DynamoDB
- SQS
- SNS
- Event Bridge

- Step Functions
- Example serverless project

Technical requirements

To experiment with AWS, you will need an AWS account with an **Identity and Access Management (IAM)** user with administrator privileges. We will try to highlight as much as possible using command-line tools rather than the cloud console, which is the graphical user interface for AWS. Our choice of programming language will be Python, except in situations where another language is mandatory. You should consult the AWS official documentation and set up the AWS CLI for the IAM user with administrative privileges.

To further automate cloud resource deployments, AWS provides a service called **CloudFormation**, which falls under the category of **Infrastructure as Code (IaC)**. CloudFormation allows AWS users to declare the cloud services and resources they need in a declarative language and execute them against a given account to deploy the services described in it. We will briefly cover CloudFormation as a refresher before the exercise at the end.

We will also make the fair assumption that you are familiar with the AWS platform and understand some of the basic constructs, especially in terms of IAM permissions, policies, and roles. Covering these topics will digress from our objectives in this chapter and hence will not be discussed.

Useful AWS concepts to know and refresh yourself on

In the following table, you can see a trimmed-down list of concepts and tools by category that you should ideally know to completely understand the project and extend it. Having a basic familiarity with AWS will be enough to understand this chapter:

IAM		
Users	Groups	Roles
Resource Policy	User Policy	Trust Relationships
Tools		
AWS Console	AWS CLI	AWS SDK
CloudFormation-cli (cfn)	SAM CLI	CloudFormation DSL
Monitoring and Housekeeping		
CloudWatch	AWS X-Ray	Tags
AWS Services		
CloudFront	RDS/Aurora Serverless	Simple Email Service

Table 3.1 – Relevant AWS concepts

In the next section, we are going to take a quick look at AWS CloudFormation – the **Infrastructure as Code (IaC)** service that will help us set up an AWS project quickly.

CloudFormation primer

CloudFormation is the AWS service that provides IaC. It allows users to describe the AWS infrastructure and services that are to be created, along with their features and configurations to be declared in a DSL. These definitions can be executed against AWS to provision the necessary resources, as described in the definition files. These files that describe IAC code are called CloudFormation templates.

We will examine the basic concepts of CloudFormation here – it is expected that you can reasonably understand CloudFormation and the rest of the AWS concepts. This section has been provided as a quick refresher only. AWS templates can be defined either in JSON or YAML format. For this book, we will use YAML-formatted CloudFormation templates as they are more readable.

Two important concepts of Cloud Formation are *template* and *stack*:

- **Template**: A text file that defines AWS resources (services and infrastructure components) in JSON or YAML format.
- **Stack**: A set of resources created from a template. These are managed as a single unit for provisioning using CloudFormation.

A template contains the following elements:

- **AWSTemplateFormatVersion**: Supported template version – the only valid value currently is 2010-09-09.
- **Description**: A textual description of what the template intends to do.
- **Mappings**: A dictionary-like data structure of key-value pairs. It can be used like a switch-case programming construct.
- **Outputs**: The values that are returned when the template is successfully executed.
- **Parameters**: Arguments you can pass to the template at runtime.
- **Resources**: This is the most important and the only required parameter in a template. Each resource defines a component that is required by the infrastructure, such as EC2 Instance, SQS Queue, ElasticSearch Instance, and so on. Each resource will have properties that define configuration options for the resource.

Now, let's look at a sample template for provisioning an AWS S3 bucket:

```
AWSTemplateFormatVersion: 2010-09-09
Description: CF Template for website S3 bucket
Resources:
 WebS3Bucket:
```

```
Type: AWS::S3::Bucket
Description: S3 Bucket via CloudFormation
Properties:
BucketName: my-personal-website
```

As you can see, this is a very simple template for creating an S3 bucket; the only property/configuration for the S3 bucket is its name. You can add more properties to the YAML block as key-value pairs. Some of these properties will be multi-level hash data structures. For example, the following block can be added under the properties section in the snippet given earlier to enable server-side encryption:

```
BucketEncryption:
  ServerSideEncryptionConfiguration:
  - ServerSideEncryptionByDefault:
  SSEAlgorithm: AES256
```

You can create this stack by saving the template to a file location and using the AWS CLI to create a stack from the command line, as follows:

```
aws cloudformation create-stack \
  --stack-name MyS3Bucket \
  --template-body file://tmp/cfntemplates/mys3.yml
```

This template does not define any parameters, but for templates that require parameters, you can pass them as command-line arguments using the `parameters` option.

Alternatively, you can upload the template to the AWS console.

Advanced constructs in the CloudFormation template language

CloudFormation is much more powerful than what we saw and can be used to construct anything in AWS programmatically. While covering CloudFormation in detail is beyond the scope of this book, I am adding a few advanced constructs for your reference. Please use them as a starting point if you would like to learn advanced CloudFormation. They are as follows:

- Resources or parameters defined in one place in the CloudFormation document can be referenced elsewhere in the template using the name of the resource.

- Use the `Map` data structure to introduce conditional logic in the template.

- Attributes of a resource are the properties of an existing resource, which can be manipulated in the template by referencing them with intrinsic functions such as `Fn:GetAtt`.

- Several built-in functions can manipulate and transform data. These functions are called intrinsic functions. To learn more, visit `https://docs.aws.amazon.com/AWSCloudFormation/latest/UserGuide/intrinsic-function-reference.html`.

When you define a stack with multiple resources, they can be updated/managed/deleted together. CloudFormation also allows you to detect drift – changes done to resources outside the stack definition – which is useful for users to resolve such conflicts or deviations.

Now, let's get started with the most important serverless service in AWS: Lambda.

Lambda – FaaS in AWS

Lambda is the serverless computing platform offered by AWS. It accepts functions and micro-applications written in one of the supported languages. These functions and applications can then be run on the Lambda platform without us having to worry about server provisioning and capacity management. As a fully managed service, all the hassles of maintaining the computing environment are abstracted away from the customer.

One of the most attractive sides of Lambda is its billing model. They follow a pricing model based on the amount of time a function runs and the memory it consumes. If you can optimize your functions and estimate the maximum time it would take to run, cost optimizing for that workload becomes easy and saves money. We will examine the pricing model in a bit more detail down the lane.

Lambda functions are completely event-driven. This means that the functions are invoked in response to an event and are run only on a need basis rather than run 24/7 like a PaaS application or a homegrown microservice would. These events could originate from any of the vast number of AWS services that are integrated with Lambda. For example, when a file is uploaded to AWS S3 object storage, a Lambda function can be invoked in response and used to solve several use cases, such as converting the file, adding additional metadata, or executing arbitrary code to post-process the file or notify another service. Lambda can also be invoked manually or scheduled to run like a cron job.

Lambda is suitable for a variety of use cases. Here are some examples:

- Processing objects in S3
- Stream processing
- APIs and serverless web applications
- IoT backends
- Mobile backends
- Event-driven architecture using SQS/SNS event triggers

We will see more use cases in the subsequent sections when we explore more serverless services that are part of AWS.

How Lambda works

Let's look at how Lambda works from inception to execution:

1. The developer creates a Lambda function using the AWS console, CLI, or SDK.

2. An IAM role is associated with Lambda to allow it access to various AWS services (at an absolute minimum, access must be provided to the logging service – CloudWatch should be given this). Lambda can create a role with minimal permissions by default.

3. A programming runtime is selected, such as Python 3.7.

4. The function code is deployed as a deployment package in one of two formats:

    - A ZIP archive containing code and dependencies

    - A container image in an OCI-compatible format

5. Once the function is ready, the function can be invoked for testing from the console or the CLIs. This invocation requires an event as input. If you are involved from the console, a test event in JSON format will be provided to you. In other modes, you need to provide an event as input.

6. When a Lambda is executed, behind the scenes, AWS provides a container with your application code and chosen runtime. It is then executed by passing an event (that triggers the invocation) and a context object to it.

7. AWS takes care of scaling the containers (based on our configurations), along with logging and metrics.

Important concepts

There are several core concepts in Lambda that every programmer should understand. They will help you in understanding the ecosystem, as well as refer to the documentation with ease:

- **Function**: A function is a FaaS construct in AWS Lambda. It encompasses everything that is required to run serverless code on top of the Lambda platform.

- **Trigger**: Triggers involve Lambda functions. These could be AWS services, which involve Lambda to respond to data or state changes happening within those services. An example of this is an item inserted into DynamoDB that triggers a Lambda function. Similarly, there are event source mappings, where messages arriving in the messaging system invoke Lambda to process them.

- **Event**: An event is the input data to a function. It's formatted as JSON and passed on to the function by the trigger services. Examples would be the details of an object uploaded to S3, an SNS notification, and a DynamoDB item that was inserted into a table. The structure of the event varies from service to service, but they all carry the data necessary to process the event.

- **Deployment package**: Code and dependencies packaged as a ZIP file or OCI container.

- **Runtimes**: Supported programming language and version combination. They support several languages, including Java, Python, Go, and JavaScript.

- **Layer**: A ZIP file containing supporting code, libraries, data, configurations, and so on. This helps in reducing the deployment package size and helps in iterating the business logic faster.

- **Extension**: These are additional programs or packages that plug into the execution runtime and have access to various events happening inside it. They use custom logic to respond to events or run processes. This is usually ideal for integrating with custom observability platforms, monitoring, and so on.

- **Qualifier**: This is the version or alias name that can be associated with a Lambda function to invoke various versions of the code.

- **Blueprints**: A blueprint is a pre-packaged Lambda function with code, configuration, and integration with selected AWS services for achieving common tasks. When you are creating a Lambda function, instead of starting from scratch, you can start with a template and modify it to suit your needs.

Important configurations

There are many configurations for a Lambda function that are useful to fine-tune its behavior. The following are some of the important ones:

- **Memory**: The amount of memory the function will use during its runtime.

- **Timeout**: The maximum time the functions will take to execute.

- **IAM role and permissions**: Configures the IAM role and associated policies to allow required permissions to access other AWS resources.

- **Environment variables**: Key-value pairs that can be set in the execution environment, but outside the code.

- **Concurrency**: The maximum number of simultaneous executions. For functions that handle a high volume of executions, reserved concurrency can help in scaling up.

- **VPC, filesystem**, and **database** proxy for connecting to the respective resources.

Lambda programming model

Let's look at the anatomy of a Lambda application and the necessary building blocks of the code.

As mentioned earlier, Lambda supports several languages and versions in the form of runtimes. Lambda's infrastructure takes care of loading the runtime and associated packages or keeps updating the runtime as and when updates and security patches arrive. There are two pieces of code that the developer has to take care of: the core Lambda code, which contains the business logic, and third-party or custom libraries that are not part of the standard runtime.

The Lambda programming model defines how your code interacts with the Lambda environment. Fundamentally, you need to define an entry point to your code from the Lambda system. This entry point is a *handler* method defined in a handler file. For example, in Python Lambda functions, the handler method and files are named `lambda_handler` and `lambda_function` by default, respectively. These can be changed while defining the function.

When the Lambda function is invoked, it receives two arguments – an event and a context. The event provides information about the AWS service event that invokes the Lambda function; for user-invoked functions, the event can be defined and supplied to the function by the user. The context, as the name suggests, shares information about the Lambda function and its environment. A handler can use both event and context data to execute its business logic. Optionally, this handler can return a result. Processing a return value and its format depends on how it was invoked. For example, a Lambda triggered by API Gateway expects the output to be in JSON serializable format.

Now that we understand the basics of Lambda, let's create a simple function that will accept two arguments – *first* and *second*. The function will calculate and return the sum of the numbers.

> **Note**
>
> Please note that I will be running all these commands from the AWS Cloud Shell. Cloud Shell is a browser-based Terminal that is pre-authenticated with our AWS credentials. To learn more about Cloud Shell, refer to `https://docs.aws.amazon.com/cloudshell/latest/userguide/welcome.html`.

```
[cloudshell-user@ip-10-4-174-161 ~]$ mkdir firstLambda
[cloudshell-user@ip-10-4-174-161 ~]$ cd firstLambda/
```

As mentioned earlier, every Lambda function needs an execution role (IAM role) to execute the function. The role will have an associated security policy. We will use the following basic policy to create our role using the AWS CLI:

```
[cloudshell-user@ip-10-2-7-250 ~]$ aws iam create-role --role-
name lambda-role --assume-role-policy-document '{"Version":
"2012-10-17","Statement": [{ "Effect": "Allow", "Principal":
{"Service": "lambda.amazonaws.com"}, "Action": "sts:AssumeRole"}]}'
....OUTPUT_TRUNCATED....
"Arn": "arn:aws:iam::756314404068:role/lambda-role",
....OUTPUT_TRUNCATED....
```

From the output, take note of the `Arn` value. This value is the unique identifier for the role you just created; we will need this while creating our Lambda function. Now, we need to associate the Lambda execution role, as follows:

```
[cloudshell-user@ip-10-2-7-250 ~]$ aws iam attach-role-policy --role-
name lambda-role --policy-arn arn:aws:iam::aws:policy/service-role/
AWSLambdaBasicExecutionRole
```

Now, let's create the function and package it into a ZIP archive. First, we will create the handler file with the `handler` function. We will use the defaults and name the file `lambda_function.py`; the handler function will be named `lambda_handler`:

```
[cloudshell-user@ip-10-4-174-161 firstLambda]$ touch lambda_function.
py
[cloudshell-user@ip-10-4-174-161 firstLambda]$ cat lambda_function.py
import json

def lambda_handler(event, context):
    sum = event['first'] + event['second']
    return json.dumps({"sum": sum})


[cloudshell-user@ip-10-4-174-161 firstLambda]$ zip first_lambda.zip
lambda_function.py
```

Now, let's create a function with the role we created and upload this ZIP archive as the code for the function:

```
[cloudshell-user@ip-10-2-7-250 firstLambda]$ aws lambda create-
function --function-name firstLambda --zip-file fileb://first_lambda.
zip --runtime python3.9 --role arn:aws:iam::756314404068:role/lambda-
role --handler lambda_function.lambda_handler
{
    "FunctionName": "firstLambda",
    "FunctionArn": "arn:aws:lambda:us-east-1:756314404068:function:fi
rstLambda",
    "Runtime": "python3.9",
...OUTPUT_TRUNCATED...
```

As you can see, the function was created with the given parameters. Since we haven't given inputs for the resources used by the function, it takes the default values for memory used and timeout – as you would see in the full output of the preceding command. Now, let's invoke the function from the command line by providing the necessary arguments:

```
[cloudshell-user@ip-10-2-7-250 firstLambda]$ aws lambda invoke
--function-name firstLambda --payload '{ "first":10, "second":20 }'
--cli-binary-format raw-in-base64-out out.json
{
    "StatusCode": 200,
    "ExecutedVersion": "$LATEST"
}
[cloudshell-user@ip-10-2-7-250 firstLambda]$ cat out.json|jq -r .
{"sum": 30}
```

As you can see, the function returned the sum of the two input numbers. This is a very simplified version of a Lambda function. To learn more about how to develop functions in Python, refer to https://docs.aws.amazon.com/lambda/latest/dg/lambda-python.html.

Now that we have a basic idea of how to create a Lambda function, let's look at a special flavor of the Lambda function – Lambda at the edge – in the next section.

Edge computing

Edge computing is a special case of Lambda and is called Lambda@Edge. As we described in *Chapter 2*, Lambda can run at the edge locations of AWS. This is made possible by the CDN service of AWS – **CloudFront**. CloudFront allows you to introduce Lambda functions in the HTTP communication path, thereby enabling users to intercept the traffic coming in and going out of CloudFront and change its behavior. Lambda@Edge is restricted in terms of capabilities and runtime support compared to its regular counterparts. This is because the Lambdas running at the edge are meant to intercept the HTTP traffic and manipulate it with minimal delays and efficiency. The following diagram should clarify the HTTP request manipulation of Lambda@Edge:

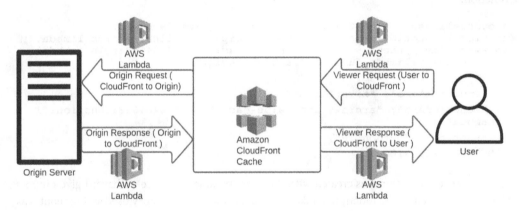

Figure 3.1 – Lambda@Edge request-response flow

Some of the use cases for Lambda@Edge include the following:

- Intelligent origin management and routing
- Search engine optimization
- Bot mitigation
- A/B testing
- Authentication and authorization
- Web security and privacy
- User metrics and analytics

In the next section, we'll look at API Gateway.

API Gateway

AWS API Gateway is an API management platform offered by AWS. It is a managed service, which means all scaling, security, and maintenance activities are taken care of by AWS. It enables developers to create, publish, and maintain APIs at any scale while providing security and monitoring. API Gateway supports both REST and WebSocket APIs, making it ideal for most of the common API needs. Some of the features provided by API Gateway are as follows:

- Scaling to handle a large number of requests
- Traffic management
- CORS support
- Security
- Authorization and access control
- Rate limiting
- Monitoring
- Versioning and life cycle management

As mentioned earlier, there are two types of APIs supported by the gateway – REST and WebSocket:

- **REST**: REST is the most common form of API and supports all sorts of functionalities that modern web applications require. These are suitable for all generic REST APIs.
- **WebSocket**: WebSockets are used for real-time communication. If you are developing a communication product, such as a chat service, a WebSocket is what you need to handle the back-and-forth communication in real time.

In the following diagram, you can see the request-response flow in AWS API Gateway:

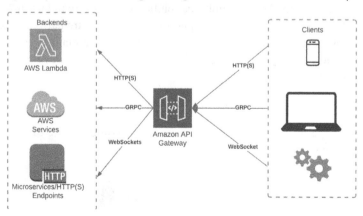

Figure 3.2 – AWS API Gateway

When you opt to create an API using the AWS console, various options will be presented. Let's take a look.

HTTP API

Here, you can build low-latency and cost-effective REST APIs with built-in features such as OIDC and OAuth2, and native CORS support. It works with Lambda and HTTP backends.

WebSocket API

You can build a WebSocket API using persistent connections for real-time use cases such as chat applications or dashboards. It works with Lambda, HTTP, and AWS services.

REST API

Here, you can develop a REST API where you gain complete control over the request and response, along with API management capabilities. It works with Lambda, HTTP, and AWS services.

Private REST API

Here, you can create a REST API that is only accessible from within a VPC. It works with Lambda, HTTP, and AWS services.

As you can see, all the options allow you to create APIs backed by AWS Lambda and selected AWS services. In the case of AWS services, it will be more like a proxy to those APIs with custom authorization and modifications to the request. Lambda can be used to build lightweight APIs in serverless mode. You can also create an API that can proxy the user request to HTTP APIs hosted either in your virtual private network within AWS or elsewhere over the public network. In addition, it allows a mock response to be returned from the gateway itself for testing purposes.

If you don't require detailed API management and want a faster and newer API implementation, choose the HTTP API option. Amazon built this solution from the ground up in 2019 and is much better in terms of performance. Amazon has plans to provide feature parity between REST APIs and HTTP APIs, but that's not materialized yet. A detailed comparison between them is provided here: `https://docs.aws.amazon.com/apigateway/latest/developerguide/http-api-vs-rest.html`.

We can deconstruct the serverless API building as follows:

1. Define the type of API you want to create – this could be one of the four categories defined previously.

2. Depending on the type, what you do next will vary. But a few things are common:

    A. **Integration**: integration defines the backends you will use for the API. This could be Lambda, HTTP endpoints, or AWS services. Depending on the type of integration, additional configurations will be provided.

    B. **Stages**: This is a construct for referring to a certain life cycle state of the API, such as dev/prod/stage. This allows developers to deploy different versions of APIs to different stages and use them from testing or segregation users.

    C. **Metrics and logging**: Monitoring your APIs can be implemented by integrating into AWS CloudWatch services.

    D. **Security**: Protecting the API can be done in a few different ways – we will touch upon them briefly later in this chapter.

    E. Both HTTP and REST need routes/resources to be defined, along with HTTP methods such as GET and POST.

    F. You can also provide a custom domain name for your API endpoint.

3. If you selected Lambda as the integration, develop and deploy the Lambda code. If the backends are HTTP endpoints, ensure they are deployed and running.

4. Once the API has been created, deploy it to a particular stage.

5. Post-deployment, test the API with some mock data – you can do this either from the console or using the AWS CLI.

With Lambda integration, API Gateway will pass the HTTP request received as an argument to the handler. This will be in the form of an event. The Lambda function should do the necessary processing of this event and form a response. The response should be in a specific format containing `statusCode` (an HTTP status code) and the body in JSON format. This object will be returned from the handler function and the gateway will pick it up and return it to the caller. Depending on your configuration, the request and response could be transformed before being passed to or returned from the Lambda function.

When creating REST APIs, there are three types of endpoints (the hostname of the API) – regional, edge optimized, and private. Geographically distributed clients can be better served with edge-optimized endpoints. Regional API endpoints are for clients in the same Region, whereas private endpoints are to be accessed from within your **Virtual Private Clouds** (**VPCs**).

API Gateway security

There are two aspects of security while working with API gateways. The first involves how the API gateway talks to other services or resources in your account. This is mostly achieved using the IAM service – that is, with security policies and trust relationships. The second and more important aspect is how the users of the APIs can be authenticated and authorized. There are multiple authentication methods supported by API Gateway, as follows:

- **Resource policies**: Here, you can apply policies to API resources to limit access from specific IPs and VPCs,

- **IAM policies and roles**: In addition to supporting API management access control, IAM can also be used to control who can invoke the APIs.

- **VPC endpoint policies**: For better control of private APIs.

- **Lambda authorizer**: This by far provides the most flexible authorization option. With this, API Gateway passes down every request to the authorizer Lambda function, at which point the Lambda function examines the request data (bearer tokens such as JWT, request-parameter information, and so on). Once the authenticity of the request has been determined, Lambda passes back a policy to the API gateway, mentioning what resource the request is allowed to access.

- **Amazon Cognito**: This provides authentication and authorization services to web and mobile apps. It also provides user sign-ups and user management with federated identity pools.

Now that we understand the concepts surrounding API Gateway, let's discuss the object store in AWS.

S3 – object store

Simple Storage Service (S3) is the object store offering provided by AWS. As we learned in the previous chapter, an object store can store unstructured data files as objects. One of the earliest offerings from AWS, S3 has come a long way in terms of features and reliability. It's used widely by AWS customers as well as AWS themselves for storing objects such as backups and archives.

S3 is highly scalable and available and adheres to high levels of security standards, all while delivering performance. S3 is used across the industry for various use cases, such as websites, video and image stores, backup and restore, IoT, big data, data lakes, and more. As per AWS, with multiple copies stored across different systems, S3 is designed with 99.99999999999% of data durability. Some of the advantages of using S3 are as follows:

- High scalability and reliability

- Extremely performant and durable

- Storage classes to suit different business use cases cost-effectively

- Security, compliance, and audit capabilities

- Query data using additional AWS services for data mining

- Automated data processing with AWS Lambda

S3 – components

As mentioned, S3 is an object storage and stores objects. AWS provides an HTTP API to manage these buckets and objects. The most important concepts of S3 are as follows:

- **Bucket**: A bucket is a namespace that is used as a container for objects. It is globally unique within all AWS S3 buckets. The S3 name should be DNS-compliant as well. You can upload any number of objects to a bucket; there is no limit. The number of buckets an account can have has a soft limit, but this can be increased by contacting customer care. Buckets are created in a selected AWS Region.

- **Object**: An object is a file with associated metadata. Metadata is an optional list of key-value pairs that provides contextual meaning to each object.

- **Object Key**: An object key is a unique identifier for an object. Since an object is contained within the bucket, this ID is also unique within the bucket. This ID could contain forward slashes and hence can look like a filesystem path with one or more directories and a filename. In reality, the entire path is the object ID.

- **Prefix**: As described earlier, you could have a forward slash in the object ID, and due to S3 having a flat structure, it will be treated as a key and not a folder structure. But despite that design, AWS understood the value of having a folder-like concept and hence introduced prefixes, which are like folders in a bucket. Prefixes allow you to create folders in a bucket and then upload files under that.

All AWS services have service endpoints. These are API endpoints that are specific to a service, be it globally or regionally. All the available S3 endpoints are listed here: `https://docs.aws.amazon. com/general/latest/gr/s3.html`.

If you are using the AWS CLI or SDKs, these endpoints will be available by default. An object in a bucket can be uniquely identified by a combination of the S3 service endpoint, bucket name, and object key. If versioning is enabled for the S3 objects, the version number should also be part of this unique URL. For example, in `https://myimages.s3.amazonaws.com/photos/profile.jpg`, *mybucket* is the bucket name, `s3.amazonaws.com` is the S3 endpoint, and `photos/profile. jpg` is the object key.

S3 bucket names follow certain standards:

- Bucket names must be between 3 and 63 characters long.
- Bucket names can consist only of lowercase letters, numbers, dots (.), and hyphens (-).
- Bucket names must begin and end with a letter or number.
- Bucket names must not be formatted as IP addresses (for example, `192.168.5.4`).
- Bucket names must not start with the `xn–` prefix.

- Bucket names must not end with the `-s3alias` suffix. This suffix is reserved for access point alias names. For more information, see `https://docs.aws.amazon.com/AmazonS3/latest/userguide/bucketnamingrules.html`.

- Bucket names must be unique within a partition. A partition is a grouping of Regions. AWS currently has three partitions: **AWS (Standard Regions)**, **AWS-CN (China Regions)**, and **AWS-US-GOV (AWS GovCloud [US] Regions)**.

- Buckets used with Amazon S3 Transfer Acceleration can't have dots (`.`) in their names.

Let's quickly touch upon a few more features of S3:

- **Storage classes**: Storage classes are S3 services with different levels of reliability, durability, and performance. The cost varies proportionally for the storage classes. A customer-facing application would use a storage class with high performance and reliability, while a backup solution would pick a storage class with slower retrieval and much lower cost.

- **S3 replication**: It is possible to replicate objects between two buckets in the same Region or different Regions. The replication could be one-way or two-way.

- **Cross-Region replication**: Used to replicate a bucket from one Region to another.

- **Same Region replication and life cycle rules**: This involves replicating a bucket to a different account in the same Region. Life cycle rules can be used to move data between different storage classes or deletions.

- **Versioning**: Versioning allows you to store multiple versions of the same object in the bucket.

- **Encryption**: Object encryption can be enabled on buckets to protect data at rest. Data in transit is always protected by TLS/HTTPS.

S3 is a very versatile service and has a lot more features to support varied use cases. S3 can act as data lakes for large amounts of data processing, can be used along with the AWS Athena service to run queries on the objects, and more.

Now, let's quickly create an S3 bucket and add content to it. Note that all bucket names should be prefixed with `s3`:

```
[cloudshell-user@ip-10-2-7-250 ~]$ aws s3 mb s3://serverless101-bucket1
make_bucket: serverless101-bucket1
[cloudshell-user@ip-10-2-7-250 ~]$
[cloudshell-user@ip-10-2-7-250 ~]$ aws s3 ls
2023-02-26 17:55:41 serverless101-bucket1
```

Let's create a dummy file and copy it to this bucket:

```
[cloudshell-user@ip-10-2-7-250 firstLambda]$ aws s3 cp s3.txt s3://
serverless101-bucket1/mypath/firstfile.txt
upload: ./s3.txt to s3://serverless101-bucket1/mypath/firstfile.txt
[cloudshell-user@ip-10-2-7-250 firstLambda]$ aws s3 ls s3://
serverless101-bucket1/mypath/
2023-03-26 17:59:20         16 firstfile.txt
```

S3 and serverless

S3 raises events when changes occur to objects. These include put, copy, delete, or life cycle events. You can make a Lambda function to respond to the selected events on matching objects. S3 will pass the details of the event – bucket name, object key, and so on – to the Lambda function, which then will take appropriate action on the event. An event notification from S3 could contain details of more than one event as an array.

S3 event notifications can also be integrated with SNS or SQS. SNS is a simple notification service, which can be used to create a fan-out architecture where an event can be sent to multiple targets, such as Lambda functions, SQS queues, APIs, and more. S3 can also deliver events to the SQS queue, which will allow the events to be processed by a non-Lambda service at its own pace. In this case, the SQS queue acts as a buffer that can create an asynchronous architecture to process the events.

S3 and Lambda are highly scalable services, and when a large number of uploads hits S3 in a short time, the Lambda function will scale up to handle the load. This scaling is constrained by the concurrency limit of the function. When this limit hits, the events are queued until an instance of a function becomes available.

To drive this usefulness, let's look at a use case given in the official AWS documentation.

Consider documents in a particular language getting uploaded to an S3 bucket. You need to translate them into another language. In a simplified architecture, how this will happen is that when a document is uploaded to S3, it sends out an event, and a Lambda is invoked. Lambda examines the document and sends it to the AWS Translate service. Translate returns the translated document, which is picked up by Lambda and then stored in another S3 bucket.

This is a simplified version of the use case. In reality, this design will be influenced by the scaling limits of Lambda and the service limits of the AWS Translation service. In scenarios where a large number of documents are being uploaded, you might have to configure S3 to send notifications to SQS first and then invoke Lambda to process the documents. Also, if the document exceeds the size limit that AWS Translated can handle, an interim Lambda function would be required to pre-process the document and chunk it. The full use case is given at https://aws.amazon.com/blogs/compute/translating-documents-at-enterprise-scale-with-serverless/.

A lesser-discussed serverless feature of S3 is its ability to serve a static website directly from storage, hence avoiding the need to have a hosting service or server for this purpose. In serverless workflows, this will come in handy when a web UI needs to be designed to work on top of an API. With JavaScript calling APIs and rendering content with S3-hosted static websites, serverless APIs can easily get a lightweight frontend.

This should have provided you with a decent understanding of S3's capabilities and how we can integrate them with serverless workflows. In the next section, we will look at DynamoDB, the versatile NoSQL platform of AWS.

DynamoDB

DynamoDB is a fully managed NoSQL service offered by AWS. It is easy to set up with minimal configurations and can scale as required with little to no effort from the developer side. It is highly performant and can provide read and write without latency. It offers security for data at rest by encrypting it. It provides high availability and durability by replicating data to multiple availability zones within a Region. Using SSDs helps it to provide fast access to the stored data as well.

Fundamentals

In this section, we are going to look into the basic building blocks or core components of DynamoDB:

- **Tables**: Just like its RDBMS counterparts, DynamoDB also stores data in tables. But unlike RDBMS relationships with foreign keys, DynamoDB does not offer any relationships, making each table an independent entity. Fundamentally, a table is a collection of data.

- **Items**: Items are like the rows in RDBMS tables. Each item identifies one unique entity of a DynamoDB table. Let's say you have a DynamoDB table to store the profile of your customers. Each item on the table would represent one unique customer. These unique items are identified by a unique key – that is, the primary key of the table.

- **Attributes**: These are equivalent to the columns in RDBMS tables. Each attribute would be a key with an associated value. In the preceding example of customer profiles, the FirstName property of the customer would be an attribute of the table and for each item in the table (one unique customer), there will be associated attributes and attribute values – say, FirstName = *John Doe*. But unlike RDBMS, not all the items in the table need to have the same attributes.

Being a NoSQL platform, DynamoDB tables are schemaless, except for the choice of having a primary key. Because of that, items in a table might not have the same attributes. Usually, the attribute values are numbers or strings. But it can also have nested attribute values – which can go up to 32 levels.

Primary keys and indexes

DynamoDB has two kinds of primary keys – simple and composite:

- **Simple primary key**: Simple primary keys contain just one selected attribute – which is termed a partition key (also referred to as a hash attribute). A hash function applied on this primary key maps the data item into a partition (internal storage abstraction of DynamoDB), hence the name partition key.

- **Composite primary key**: Composite primary keys contain two attributes. The first one is called the partition key, as before, and it has to be unique across all the items in the table. The second attribute is a sort key, which could be any other attribute of the table. A partition key is used to partition the data while all the items in the same partition are sorted in the order of the sort key value. In this case, multiple items can have the same partition key. The sort key is also known as the range attribute.

All primary key attributes should be scalar – number, string, or binary. Similar to RDBMS, primary keys are used for indexing the table data for faster querying. A table will always have a primary index based on the primary key. Since it is mandatory to have a primary key, the primary index comes along with it. But sometimes, having a primary index is not enough. Depending on the data stored in the tables, the business use cases might require querying based on other attributes of the table.

Secondary indexes allow developers to query the table data using an alternate (non-primary) key. A secondary index built on an attribute can be treated just like the table (which is also the primary index). After creating the secondary index, you can query the attributes against that index. You can create more than one secondary index on a table. In the context of secondary indexes, the original table on which the index is created is referred to as the base table.

There are two types of secondary indexes in DynamoDB – global and local:

- **Global secondary index**: This is an index with a partition key and sort key that can be different from those on the table. The maximum number of global secondary indexes that can be created on a table is 20.

- **Local secondary index**: This is an index that has the same partition key as the table, but a different sort key. The number of local secondary indexes per table is limited to 5.

For more information, you can check out the *General Guidelines for Secondary Indexes in DynamoDB* section here: https://docs.aws.amazon.com/amazondynamodb/latest/developerguide/dynamodb-dg.pdf.

There are three key terms when interacting with DynamoDB data:

- **Query**: All read and write operations on a table item

- **Scan**: An operation that's performed on an entire table or part of it (similar to the variants of the SQL select * query)

- **Filter**: Rules that can be applied to the result of an operation – query or scan – before it is returned to the user/application performing the operation

As mentioned earlier, DynamoDB is highly performant and scalable, with little in the way of server management. The following are some of the important features that make DynamoDB a fantastic cloud NoSQL platform:

- Global tables for multi-Region replication

- Using Time-To-Live to auto-expire items

- On-demand backup

- Point-in-time recovery

- Caching layer to speed up access using DAX, DynamoDB's accelerator

- Two capacity models to process reads and writes – on-demand and provisioned

- Two data read consistency models – eventual consistency and strong consistency

- Integration with other AWS services for data processing; for example, for exporting data to Amazon Redshift for data analysis

- Authentication to access individual items using Cognito user pools

DynamoDB and serverless

AWS Lambda is at the center of most serverless processing and computation. DynamoDB has introduced a tight integration with Lambda using DynamoDB Streams.

DynamoDB Streams falls under the category of **Change Data Capture** (CDC) systems. A simple definition of CDC systems is that they capture the changes occurring in a data system and relay them to interested parties. AWS introduces DynamoDB Streams as follows:

> *"DynamoDB Streams is an optional feature that captures data modification events in DynamoDB tables. The data about these events appear in the stream in near-real time, and in the order that the events occurred."*

Each event in DynamoDB Streams is called a record, which is a snapshot of a table item that is newly added, modified, or deleted. In the case of modification, the before and after snapshots of the item are included in the record. Each record also has useful metadata such as table name, change timestamp, and more, and has a lifetime of 24 hours.

DynamoDB Streams can integrate with Lambda. It can act like a trigger – a function that will launch every time a new record is added to the DynamoDB stream. This will allow developers to build a lot of business workflows around streams. For example, consider a user shopping on your website; when the record of the sale is inserted into DynamoDB, an email can be sent to the user to inform them about the details or confirm the sale.

You will see a live example of creating a DynamoDB table, loading data into it, and using it from Lambda in the project that is provided at the end of this chapter. In the next section, we are going to learn about **Simple Queue Service (SQS)** which is one of the leading message queue services in AWS.

SQS

Amazon SQS is a powerful managed message queue service. It is very secure, durable, and highly available. As we discussed in *Chapter 2*, message queues are systems that help distributed systems decouple their constituent software systems. Being accessible over generic web service APIs that you can integrate with using standard AWS SDKs makes building distributed systems around it much easier. Developers use SQS to scale and decouple microservices, distributed systems, and serverless applications.

SQS enables the users to send, receive, and store messages between dependent software components in an asynchronous fashion. It can handle large volumes of data without scalability challenges or data loss.

Types of queues

There are two types of queues in SQS – standard and **first-in, first-out (FIFO)**. They differ in terms of their ordering and delivery guarantee, as well as scaling:

- **Standard queue**: This has unlimited throughput, an at-least-once-delivery, guarantee, and best-effort-based message ordering. Messages can sometimes arrive out of order. Duplicate messages could be produced occasionally. Standard queues are suitable for most high-volume use cases. Applications handling real-time user request processing would benefit from the high throughput standard queue provides. A request can be accepted at a high rate while the backend processing can be asynchronous.

- **First-in, first-out queues**: As the name suggests, the ordering of messages is what distinguishes FIFO queues from standard queues. FIFO preserves the order of messages received. This is critical in situations where events can be processed out of order. Consider the case of a multi-step workflow where online shopping is processed. Without ordering, the product could get shipped before the money is deducted from the customer's account. With FIFO, there are no duplicates – it guarantees exactly-once processing. While FIFO may not scale at the same scale as a standard queue, it can still handle a high number of operations per second, and the performance improves much more if the operations can be batched.

Features

Some of the important features of SQS are as follows:

- A **dead letter queue** is for re-processing or discarding events that failed to be processed the first time probably due to being oversized, in the wrong format, and so on.

- Messages within the SQS queue can be secured by using **Server-Side Encryption** (**SSE**). AWS Key Management Service can store encryption keys, which can be used to apply SSE.

- Larger messages that exceed the SQS message size limit can be stored in S3 or DynamoDB and SQS will transparently handle the processing of such messages.

- SQS uses a locking mechanism during message processing to ensure multiple producers can send and multiple consumers can consume messages at the same time without any conflict.

- Durability, availability, and scalability are key features of SQS. It employs redundant infrastructure, request buffering, ordering and delivery guarantees, and more to do so.

The life cycle of a message in SQS is as follows:

1. The producer sends a message to the SQS queue.

2. The consumer picks up the message from the queue.

3. The queue starts a visibility timeout (this is configurable – it could be from zero seconds to 12 hours).

4. During the visibility timeout, the message remains in the queue but is invisible to other consumers.

5. If the consumer processes the message and deletes it within the visibility timeout, that's fine.

6. If the consumer fails to process the message, SQS will make it visible after the visibility timeout, at which point any active consumer can claim it.

All messages have a retention period, after which they will be deleted from the queue. The maximum retention period is 14 days.

You can create an SQS queue, then send and receive a message with the AWS CLI, as follows:

```
[cloudshell-user@ip-10-2-7-250 firstLambda]$ aws sqs create-queue -
queue-name serverless101-q101
{
    "QueueUrl": "https://sqs.us-east-1.amazonaws.com/756314404068/
serverless101-q101"
}
[cloudshell-user@ip-10-2-7-250 firstLambda]$ aws sqs send-message
-queue-url https://sqs.us-east-1.amazonaws.com/756314404068/
serverless101-q101 --message-body='{"mymsg":"hello sqs"}'
{
    "MD5OfMessageBody": "7dace606e6a12623ad381d473f7734b7",
    "MessageId": "2c182cfe-4cc5-40aa-9a32-06238cca6e6e"
}
[cloudshell-user@ip-10-2-7-250 firstLambda]$ aws sqs receive-
message -queue-url https://sqs.us-east-1.amazonaws.com/756314404068/
serverless101-q101
...OUTPUT_TRUNCATED...
        "  "Messag"Id": "2c182cfe-4cc5-40aa-9a32-06238cca6"6e",
```

```
...OUTPUT_TRUNCATED...
            "Body": "{\"mymsg\":\"hello sqs\"}"
...OUTPUT_TRUNCATED...
```

As you can see, we were able to create a queue, then send and receive a message. The only key parameter while using an SQS queue is its URL, which you can obtain while creating the queue, or later by getting the queue information.

SQS and other AWS messaging products

Amazon SQS and **Simple Notification Service** (**SNS**) are managed queue and topic services, respectively, which are highly scalable and don't require any server management. They have high scalability and simple APIs.

Amazon MQ is a managed message broker service that provides compatibility with popular message brokers and protocols. It is useful when migrating existing messaging stacks and the customer wants to preserve compatibility. Supported brokers/protocols include JMS, AMQP, MQTT, OpenWire, and STOMP.

SQS and serverless

SQS's scalability and minimal configuration and management make it an ideal choice for serverless messaging. SQS has very good integration with Lambda. Users can use Lambda to process messages in the SQS queue. The Lambda platform polls the SQS queue and invokes the associated Lambda function when messages are received in the queue. This is highly scalable and makes it easy to process data asynchronously. These messages could be batched for faster processing. Depending on the number of messages, Lambda will use more instances of the Lambda function to process the excess requests. If Lambda fails to process the message for some reason, it can be configured to send the messages to a dead letter queue.

To configure the Lambda processing of SQS messages, a trigger should be added to the intended Lambda function. Lambda supports multiple triggers, with SQS being one of them. There are configuration parameters to define batch sizes or time to wait before batching available messages. This process is also called event source mapping. Depending on the messages and batch size, the Lambda timeout might have to be increased. If you're expecting large volumes of messages, having reserved concurrency for the Lambda function will help.

In the next section, we will explore SNS, a much more versatile AWS service for messaging.

SNS

AWS SNS provides managed message delivery services between publishers and subscribers. Due to this characteristic, it falls into the category of messaging systems that we commonly refer to as pub-sub messaging – for publisher-subscriber. These publishers and subscribers are otherwise known as producers and consumers, and you might see these words used interchangeably in the documentation. The communication between publishers and subscribers is asynchronous – as we explained in the first part of this book. This allows for designing applications that are decoupled and can be scaled independently.

SNS provides a topic – which AWS refers to as a communication channel – that is the central piece of the communication system. SNS publishers – usually an application or microservice – push messages to the topic. Subscribers – which can be other AWS services or microservice applications – can subscribe to the topic and receive notifications (messages).

There are two categories of subscribers in this communication model – **Application to Application (A2A)** and **Application to Person (A2P)**. Subscribers are sometimes referred to as SNS targets, while the publishers are referred to as event sources.

In A2A, the consumers of the notifications will be an AWS service or a microservice application. Some of the endpoints include the following:

- AWS Lambda
- AWS SQS
- AWS Kinesis Data Firehose
- HTTP/HTTPS endpoints

In A2P, the consumers of this category of notification will be mobile application users, email users, and phone users. The notifications will be delivered to the following endpoints:

- Android/Apple push notifications
- Emails
- SMS

The following diagram shows the data flow between the publishers and subscribers of the SNS service:

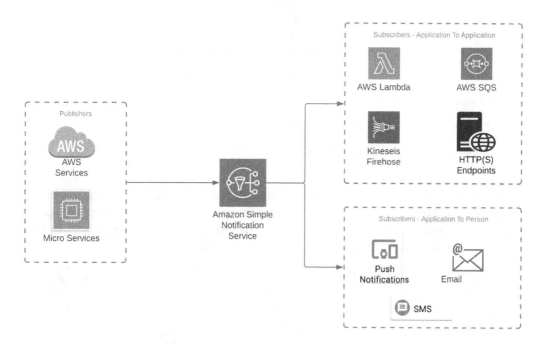

Figure 3.3 – AWS SNS publishers and subscribers

Like SQS, SNS also has two flavors of topics – standard and FIFO. FIFO ensures ordering; the only subscriber allowed for a FIFO topic is an SQS FIFO queue.

After creating an SNS topic, at least one subscriber should subscribe to this topic before the producers start posting messages. When creating a subscriber, the fundamental configuration that is needed is the protocol and endpoint. The protocol is the type of destination/subscribers supported by SNS – as you saw when we described the A2A and A2P models. The endpoint is the final destination that matches the protocol type. For instance, if the protocol chosen is AWS SQS, the ARN of the SQS queue will be the endpoint. If the protocol is email, one or more email IDs would be the endpoint(s). Depending on the protocol and endpoint, there will be more configuration parameters to fill in.

There are two optional but oftentimes required steps for high traffic and more complicated business use cases. They are filter policy and dead letter queue. Filter policy allows a subscriber to receive a subset of notifications based on some matching rules. These policies are based on the values of the notification attributes. A dead letter queue is an SQS queue that provides a temporary holding space for messages that SNS failed to deliver to its subscribers. Subscribers can reactively inspect these queues and pick up the failed messages for processing again.

Some of the additional features of SNS are as follows:

- Reliable storage of messages across geographically dispersed servers and storage
- A retry policy to ensure delivery to endpoints that failed to receive the messages
- Dead letter queues and filter policies
- Add message attributes for additional context
- SSE to ensure message security while at rest

The primary advantage of SNS is the fanout architecture it provides. Fanout is the process of distributing a message from a single source to several destinations, all of which can process the messages independently. This is a powerful concept as it liberates the producer from worrying about who is picking up its messages. This also adds the flexibility to add more subscribers to a topic without the producer ever having to worry about it:

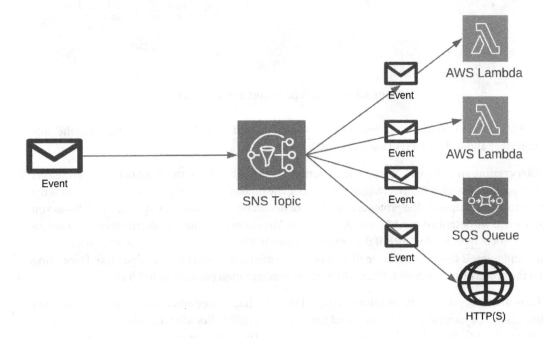

Figure 3.4 – SNS message fan-out

As you can see in the preceding diagram, the same event is pushed to multiple consumers. This is the fan-out architecture and it scales quite well.

SNS and serverless

As discussed earlier, the biggest advantage of SNS is its fan-out capability. A message received in an SNS topic can be sent to different types of subscribers and at a huge scale. This enables developers to build automations with minimal to zero coding. Consider the case where an event that occurs in your infrastructure has to be sent to both your production and staging environments. The SNS topic can have two SQS queues as subscribers, one each for production and staging. SQS can hold the events for a long time and computing services (including but not limited to Lambda) can pick up these events and process them. This provides a powerful serverless platform for building complicated pipelines.

The most important serverless feature is integration with Lambda. It allows a Lambda function to be triggered for every event that is received by the SNS. Lambda will receive the published message as a payload – into its handler function as a payload, which can subsequently be processed by it. While the workflow is simple, it couples the pub/sub platform with the serverless compute platform and can be used as a self-sustaining event-processing environment.

Another advantage of SNS is the fact that a large number of AWS services can publish events to SNS topics without involving any additional services or code. For example, when an EC2 auto-scaling group scales the number of instances to accommodate the increasing traffic, it can send a notification to a configured SNS topic. This topic can have an SMS subscriber alert the on-call engineer whenever the event happens, which will help the operators of the autoscaling infrastructure keep an eye on how their stack is scaling (or failing to scale).

Several such AWS services can be publishers to the SNS topic, thus creating an event-processing architecture that allows developers to respond to changes that are being made to their services and data. These services are also called event sources of SNS. To find the complete list, please visit `https://docs.aws.amazon.com/sns/latest/dg/sns-event-sources.html`.

As you can see, the fan-out architecture provided by SNS is very powerful. With a large number of event sources and destinations, SNS can mediate between a multitude of systems, building an event-driven architecture with minimal code. This makes it a centerpiece in the serverless ecosystem.

In the next section, we are going to take a quick look at EventBridge, an AWS service in the category of enterprise event buses.

AWS EventBridge

AWS EventBridge is a publisher-subscriber service that falls into the category of enterprise event buses. It evolved from the AWS service CloudWatch Events. Event buses are usually used to propagate messages between a small number of internal services in a microservice or service-oriented architecture. As in SNS, the services sending messages are called event sources, while the recipients are called targets. At the time of writing, EventBridge supports 130 event sources and 35 targets. The sources and targets could be AWS services, custom services/endpoints, or SaaS partners. The integration with partner SaaS platforms such as Pagerduty, Shopify, Auth0, and others are unique to EventBridge compared to SNS and SQS.

Like SQS queues and SNS topics, the medium that stores events in EventBridge is called an event bus. But unlike SNS or SQS, you don't provision event buses per application. The default event bus receives and disperses all events from AWS while the partner event bus manages events from SaaS partners. This makes the event bridge buses more serverless and relieves individual producers and consumers from having to worry about provisioning or maintaining the bus. In terms of message guarantees and delivery, event buses act similarly to SNS standard topics. The order of messages is not guaranteed. Message delivery is guaranteed to be at least once.

Like SNS, EventBridge also provides fan-out capability, albeit with limitations. An event can only be fanned out to five target services. The scale of events that can be handled is also way less compared to SNS at the time of writing.

Now that we have an understanding of the basic features of EventBridge, let's look at how the data flow works:

- An event source publishes an event to the event bus. Depending on the source of the event and your application design, this could be your default event bus, partner event bus, or custom event bus.

- Users can create rules to match certain events based on their content, and then route the events to one or more targets (not more than five, though).

- The target(s) accepts the events and acts on them.

Some of the important features of EventBridge are as follows:

- **Scheduled events**: Scheduling allows users to implement timed delivery of events to target services. This scheduling will be based on a cron-like expression that can be configured as part of the routing and filtering rule.

- **Schema registry**: One of the biggest challenges in event-driven architecture is identifying the structure of the event. EventBridge solves this by introducing a schema registry, which can ingest event samples from an associated event bus and auto-generate schemas.

- **Archive and replay**: Customers can store events in an event archive and can be readmitted to the event bus as needed. This is useful in troubleshooting and examining historical data.

- **Message filtering based on content**: SNS could filter events, but only based on the message attributes (a maximum of 10 attribute matches). EventBridge allows filtering based on the contents of the message. Events are in the form of a JSON document and filtering with rules allows the customers to specify matches based on the JSON data types in the message schema.

- **Input transformation**: This feature allows the event structure to be transformed before it is passed on to target services. This eliminates the need to have a glues service or code that will do the transformation programmatically. This improves the serverlessness of EventBridge.

- In addition, it has all the same common reliability and security features that other AWS messaging services have – multiple copies of data stored in independent locations, server-side encryption for data at rest, TLS encryption for data in transit, and more.

Now that we have covered the core messaging services in AWS, let's explore the workflow automation service provided by AWS.

Step Functions

Step Functions is an orchestration service provided by AWS to automate interactions between various AWS services, including Lambda, in a customer-defined sequence to build complicated business workflows. Every business process or task can be defined as a set of atomic steps that execute a specific piece of work. Each of these steps has inputs and outputs; the output of one step could be fed as the input to the next step. After a particular step, the following steps could be done in sequence or parallel, or even branch out, depending on certain conditions.

This process is what we call a workflow – it can also be called a state machine. The fundamental idea behind calling a workflow a state machine is that the workflow (or the business process it represents) has a specific state at any given point in time. As it progresses through the workflow, it changes from one state to another (called transition) until it reaches its final state. Let's clarify this with a simple example – consider the case of a customer placing an online order:

1. In the initial stage of purchase, the item is in the cart and the order is in the *initiated* stage.
2. Once the purchase proceeds to the next stage, the application will internally check if the order is *deliverable* or not – this becomes the next stage.
3. If not deliverable, the workflow ends in a *failed* state.
4. If deliverable, the state is *deliverable* and the process moves to payment processing.
5. If payment fails, the process state also changes accordingly, subsequently leading to the workflow ending in a failed state.
6. If the payment succeeds, the order goes into the succeeded state.

In a real-world scenario, this process will move further into shipping and delivering products ordered, cancelations, returns, and more, creating a more complicated workflow with a large number of states and transitions.

AWS Step Functions provides a serverless workflow service and can integrate with multiple AWS services for each step of its workflow. The billing model for Step Functions is based on the number of state transitions a workflow performs during the execution. For defining a workflow, AWS provides a JSON-based **Domain-Specific Language** (**DSL**) to define and interconnect states – called **Amazon State Language** (**ASL**). In addition to ASL, AWS also provides a graphical UI via the AWS console that visually creates and manages the workflow. This tool is called Workflow Studio.

There are several states that Step Functions supports. They are as follows:

- **Task state**: A unit of work under execution
- **Choice**: Deciding how to branch off based on the input
- **Fail/Succeed**: Stops the execution depending on the outcome of workflow execution
- **Pass**: Passes on input data to output data, or inserts some additional data
- **Wait**: Introduces delays
- **Parallel**: Branches parallelly to execute the same task multiple times
- **Map**: Dynamically iterate steps

There are two flavors of Step Functions – express and standard. They differ in terms of reliability, auditability, scale, and integration. Standard workflows are suitable for long-running and auditable workflows – they can last a year, and the execution history can be retrieved for 3 months. It guarantees utmost-once execution and is ideal for non-repeatable workflows, which can cause problems if they're executed again and again. Express workflows, on the other hand, can only last 5 minutes and CloudWatch logging is the only way to inspect the execution. They are ideal for high-volume event processing and have higher limits for executions. There are two types of express workflows – synchronous and asynchronous, which provide at-least-once and at-most-once execution guarantees, respectively.

Some of the Step Functions features that make it such a robust workflow execution engines are as follows:

- **Control input and output**: A step function accepts input in JSON format, which can be further processed at three stages within that step – the input path to filter the input further, the result path to select the combination of input and task data to pass to the output data, and the output path, where the result path data can be further filtered.
- **Error handling**: ASL allows users to define error catching and retries for gracefully handling errors.
- **AWS service integrations**: AWS services can be integrated as tasks, called service tasks. Several computer, database, messaging, and other services can be used as steps in the workflow.
- **Integrate with APIs**: Step Functions can integrate with HTTP(S) endpoints running anywhere to build more robust workflows.
- High availability, security, compliance, and scalability.

The following is a snapshot of the visual representation of Step Functions from Workflow Studio. The image was taken from the AWS documentation:

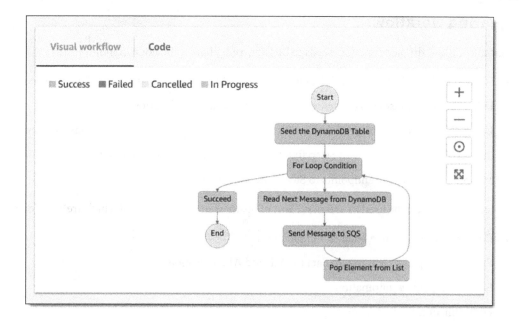

Figure 3.5 – Step Functions in Workflow Studio

Now, let's look at the various service integrations supported by Step Functions.

AWS service integrations

Step Functions can invoke other AWS services in each of their steps. There are two ways to do this – optimized integrations and AWS SDK integrations. SDK interactions allow Step Functions to call any AWS APIs from within the state machine. Optimized integrations are the recommended way to invoke AWS services whenever possible. Some of the optimized integrations are as follows:

- Invoke an AWS Lambda function
- Add or retrieve an item from a DynamoDB table
- Run an Amazon ECS/Fargate task
- Publish a topic in SNS or queue in SQS
- Launch another step function

The complete list of integrations can be found at https://docs.aws.amazon.com/step-functions/latest/dg/connect-to-services.html.

Triggering workflows

To start a workflow, the step function needs to be triggered. This can be done in multiple ways:

- Using the AWS API to make a Call StartExecution API call from the AWS Step Functions API

- Using the AWS console to trigger a new execution of the step function

- Using AWS CloudWatch events or EventBridge to trigger Step Functions in response to events

- Triggering the execution of one step function from a step in another step function

- Mapping API Gateway endpoints to Step Functions APIs

AWS Step Functions can be used to automate a lot of business use cases, some of which are as follow:

- Data **extract-transform-load** (ETL) pipelines

- Pre-processing and cleaning up data for ML and AI applications

- Business processes automation

- IT automation

- Security and vulnerability scanning and response

- Image and document processing

This is not an exclusive list, but you get the idea. Step Functions reduces the glue code that needs to be written for connecting various parts of a business process and sometimes even eliminates it. It can replace a lot of the business logic and even extend the capabilities of API Gateway with integrated workflows triggered for API calls.

Doing by example – lightweight alert manager

Service monitoring and alerting is an essential practice for any infrastructure. We use a lot of tools to collect active and passive check results, metrics, and more. Once a problem has been detected, it is essential to alert the owner of the broken system. The solution we are designing is for alerting for an issue that's been detected in any monitoring systems. These issues are called incidents, and they need to be acknowledged by an on-call engineer, who is responsible for managing a service's infrastructure. If the on-call fails to acknowledge the incident (and subsequently work on it), the issue should be escalated to the team manager or other leadership, depending on the escalation policy defined.

High-level solution design

The following diagram summarizes the implementation:

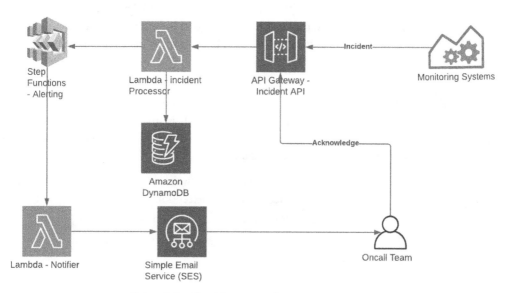

Figure 3.6 – Alert Manager architecture diagram

Let's take a quick look at how this works.

The monitoring system will post any incidents to the API gateway, which is responsible for creating, updating, and retrieving incident data. This logic will be implemented in the incident processor Lambda function. When a new incident is created, Lambda inserts that incident into the DynamoDB table for incidents. Then, Lambda will look up the incident details and find the owning team and its on-call and escalation matrix from the database. This escalation and incident information will be passed to a step function asynchronously. The step function will use the notifier Lambda function to send an email to the primary on-call immediately (using **Simple Email Service** (**SES**)). It will wait for the escalation time (by default, this is 10 minutes) and then check the status of the incident. If the incident is not acknowledged by the on-call (who will hit the API gateway to acknowledge the incident), the escalation email will be sent to the email listed in the escalation matrix.

This should provide you with enough ideas about the service components. Now, let's build the solution. All provisioning will be done with CloudFormation; the Lambda code will be in Python. The step function will have a DSL-based definition.

The project, along with all the code and automation, is available at `https://github.com/PacktPublishing/Architecting-Cloud-Native-Serverless-Solutions/main/chapter-3`. The setup instructions are attached; you should be able to get up and running in less than 5 minutes.

Summary

AWS is the most advanced cloud provider we have currently, and they are good at pushing new services fast onto the market. Lambda and the serverless ecosystem have grown multifold and have become synonymous with serverless and FaaS. The current AWS focus, in terms of Lambda, seems to be toward performance improvement as well as more integration with other AWS services. This will open up more possibilities for serverless applications and newer architecture patterns that will help process data easily and seamlessly.

In this chapter, we learned about the core AWS serverless services, with Lambda being at the center. We looked at how these services can be integrated and how they can help in designing complex serverless architectures. As AWS is the cloud provider that set a standard for serverless, this chapter was vital as it helped you easily understand the upcoming chapters.

The next chapter is on serverless platforms in Azure and for the most part, Azure services can be mapped one to one with AWS serverless services with a few exceptions.

Serverless Solutions in Azure

Azure was a late entrant to the serverless revolution, but it has been quickly catching up with AWS in terms of services and feature parity. Just as the Google GCP cloud comes with the experience of running Google Search and Workspace at scale, Azure comes with a track record of running Bing and Office 365 at a global scale. Bing and Office 365 also bring a number of productivity and enterprise applications into the fold, many of which can be integrated in some way with the Azure serverless offerings.

In this chapter, we will learn about the core serverless services that Azure offers. We will kick off by looking at some fundamentals of Azure and then warm up with an introduction to Azure Functions, followed by other key services in the serverless portfolio. The idea is to give you enough foundational knowledge to start experimenting with Azure serverless and then give you pointers to research further and learn on your own.

We will cover the following topics in this chapter:

- Azure fundamentals
- Azure Functions
- Azure Blob Storage
- Azure Cosmos DB
- Azure event and messaging services
- Azure Logic Apps
- Project – image resizing

Technical requirements

You will need some basic skills and cloud access to effectively understand this chapter. The following are the minimum requirements:

- An Azure cloud account. You can sign up for a free account here: `https://azure.microsoft.com/en-in/free/`.

- After signing up, you should be able to log in to the Azure cloud console (the Azure portal).

- Most of the examples given in this chapter will be done either from the Azure portal or from the command line.

- The development machine, laptop, or server should have a shell – preferably one of bash, sh, or zsh – which are usually the defaults on most Linux, macOS, and **Windows Subsystem for Linux (Windows WSL)** distributions.

- Install the Azure CLI as described at `https://docs.microsoft.com/en-us/cli/azure/` on your developer machine.

- Once the CLI is installed, authenticate your CLI with the following: `https://docs.microsoft.com/en-us/cli/azure/authenticate-azure-cli`.

- It is recommended to have the Azure Functions core tools installed as well. Follow the instructions here: `https://github.com/Azure/azure-functions-core-tools`.

- We will also be using the JSON processor CLI – jq – for a lot of output manipulation. Install your platform-specific version by following the instructions here: `https://stedolan.github.io/jq/`.

Besides these, we will make a few technological choices in the chapter to streamline the learning:

- Whenever possible, we will choose the command line for creating and managing cloud entities.

- When programming is required, Python will be chosen wherever possible. This is just an arbitrary choice, based on the assumption that a reasonable majority of our readers will be familiar with it.

- We will keep all the code examples as simple as possible, using minimal programming constructs when necessary. This will help readers who are new to Python or don't know it.

- While we have chosen Python, the examples in this chapter can be written in any programming language supported by the respective Azure platforms.

- If a configuration supports both YAML and JSON, we will gravitate toward YAML for better readability.

- For all demonstrations in this chapter that require a command line rather than a developer machine, I will be using Azure Cloud Shell. Follow the instructions here to launch Cloud Shell: `https://learn.microsoft.com/en-us/azure/cloud-shell/overview`.

Azure fundamentals

As mentioned earlier, in order to start using Azure, you need an Azure account. There are a few fundamental concepts that you need to know about to better manage cloud entities in Azure. One of them is the organizational constructs used in Azure to manage cloud resources. Let us first take a look at the following diagram and then learn about each of the elements in the hierarchy:

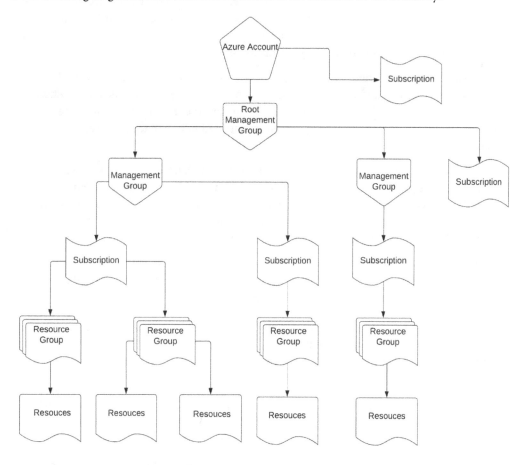

Figure 4.1 – Azure resource hierarchy

Let us understand each item in this hierarchy. Note that the descriptions are not necessarily in the order of the hierarchy. This is to make describing the relationships a bit easier:

- **Azure Account**: The top-level entity in Azure, an Azure account contains everything about your Azure cloud access, including all cloud assets, agreements, billings, and so on.

- **Resources**: These are instances of Azure cloud services that you can create. Some examples of resources are virtual machines, databases, object store buckets, and so on. They are the basic unit of cloud assets and can't be subdivided further.

- **Resource Groups**: These are containers within which you can create resources. Ideally, a resource group contains related Azure resources – such as all VMs and DBs used to run a particular web application in a developer environment or production. They are also the lowest level of grouping that can be done in Azure. It is easy to apply access control to resource groups as the resources within the group inherit those controls.

- **Subscription**: A subscription is a grouping construct for cloud assets in your account. It logically associates the account and the resources created in it. Subscriptions have defined quotas and limits on the number of assets or the amount of data or other features that can be used. They can be used for managing and designing your costs and billing. Each resource can only be deployed to one subscription.

- **Management Group**: The management group is the top-level container in the Azure organizational hierarchy. It acts as a container for subscriptions and other management groups. You can assign policies, access control, and other management constructs to management groups, which will be inherited by all subscriptions and management groups under them, making it easy to implement governance on top of your Azure resources.

- **Root Management Group**: A root management group is the default management group that is created within your account. This group can contain other management groups and resources. Unlike other management groups, root management groups can't be deleted.

Management groups and subscriptions allow you to model your cloud strategy and mimic your organization hierarchy in your cloud account as well. This helps with better management of billing and accounting. It also allows you to enforce policies and access control in a centralized and manageable way.

While we have all these facilities to organize our resources, in the scope of this chapter, we may not need such elaborate resource organization. If you are just starting with your free account, you will have one subscription and all resources can be created under that subscription. This subscription defines the limit of free cloud resources that you can create under the free account.

One other concept that is useful in managing your cloud account is location. Location is a synonym for the region, which is a group of Azure data centers located close to each other within a geographic location. This allows Azure customers to deploy resources in a location that is close to their customers and businesses.

The availability zone is a concept that is closely connected to regions. Availability zones are physically separate locations within an Azure region. This could be one or more data centers that are very closely interconnected with low or close to zero latency between them. Local failures such as power and network outages or even natural calamities are usually contained within an availability zone, leaving other availability zones within the region to operate normally.

It is always recommended to set a default location or region in your CLI or the portal. You can always switch this per command invocation or in the portal for any interaction. To do this at the CLI, first list the locations and find your preferred region. I am looking for regions in India that are closest to my location:

```
safeer [ ~ ]$ az account list-locations|jq -r '.[]|select(.
displayName|contains("India"))|[.name,.displayName,.metadata.
physicalLocation]|@csv'
"centralindia","Central India","Pune"
"india","India",
"jioindiawest","Jio India West","Jamnagar"
"jioindiacentral","Jio India Central","Nagpur"
"southindia","South India","Chennai"
"westindia","West India","Mumbai"
```

As you can see, there are multiple regions in India. Since I am located in South India, the Chennai location – named `southindia` – would be closest to me. I will set that for my CLI here:

```
safeer [ ~ ]$ az configure --defaults location=southindia
safeer [ ~ ]$ az configure -l
[
  {
    "name": "location",
    "source": "/home/safeer/.azure/config",
    "value": "southindia"
  }
]
```

It is recommended to create all resources under one resource group for ease of management and cleanup after we are done with those resources. Let us create this group using the Azure CLI:

```
safeer [ ~ ]$ az group create --name serverless101-rg

{
  "id": "/subscriptions/7u6f9085-h345-4bnh-aa54-b030af67871e/
resourceGroups/serverless101-rg",
  "location": "southindia",
  "managedBy": null,
  "name": "serverless101-rg",
```

```
  "properties": {
    "provisioningState": "Succeeded"
  },
  "tags": null,
  "type": "Microsoft.Resources/resourceGroups"
}
```

Now, let us set this as the default resource group so that all resources we create going forward will be under this group (and you do not have to explicitly set the group in every command):

```
safeer [ ~ ]$ az configure --defaults group=serverless101-rg
safeer [ ~ ]$ az configure -l|jq '.[]|select(.name=="group")|[.value]'
[
  "serverless101-rg"
]
```

Another entity that is required to manage Azure is a storage account. An Azure storage account is a container for different types of data storage services. It provides you with a unique namespace to organize your data from different services as a group. Within the storage account, you can have storage entities from the following services:

- Blob Storage

- Azure Files

- Queue Storage

- Static websites

- Table storage

- Data Lake Storage Gen2

There are different types of storage accounts. They offer different storage features and the pricing model depends on the type. For our use cases, we will use Standard general-purpose v2. To learn more about storage accounts, visit https://learn.microsoft.com/en-us/azure/storage/common/storage-account-overview.

Now, let us create a storage account as follows. Note that storage accounts need a unique global name. We will also provide the SKU (type of storage) name Standard_LRS – locally redundant storage. To learn more about the SKU ref, you can visit https://learn.microsoft.com/en-us/rest/api/storagerp/srp_sku_types:

```
safeer [ ~ ]$ az storage account create --name serverless101storage1
--sku Standard_LRS
{
  "accessTier": "Hot",
 OUTPUT_TRUNCATED_AY_MULTIPLE_PLACES
```

```
  },
  "primaryEndpoints": {
    "blob": "https://serverless101storage1.blob.core.windows.net/",
    "dfs": "https://serverless101storage1.dfs.core.windows.net/",
    "file": "https://serverless101storage1.file.core.windows.net/",
    "queue": "https://serverless101storage1.queue.core.windows.net/",
    "table": "https://serverless101storage1.table.core.windows.net/",
    "web": "https://serverless101storage1.z30.web.core.windows.net/"
  },
  "primaryLocation": "southindia",
   "resourceGroup": "serverless101-rg",
  "sku": {
    "name": "Standard_LRS",
    "tier": "Standard"
  },
```

As you can see from the truncated output, the storage account provides the endpoint URL for all supported services, and it took the default location and resource group that we set earlier.

Now that we have created all the prerequisites, let us look at our first Azure serverless service, which obviously will be the FaaS platform of Azure: **Azure Functions**.

Azure Functions

Azure follows the same FaaS platform principles as other vendors do. It is a functions platform where you can trigger function code with events or schedules. There is no need to provision any infrastructure. It is a production-quality service that can be used for event-driven services, APIs, WebHooks, and many more such use cases. It has pay-as-you-go billing and scales automatically.

A couple of core concepts of functions that you need to learn about first are function triggers and bindings. A trigger is what invokes a function, and a function can only have one trigger. Triggers can supply inputs to a function if the triggers themselves have associated payloads. An HTTP trigger is a trigger that comes via an HTTP endpoint and invokes the function. In this case, the HTTP request will become the input payload for the function.

Bindings are the declarative way to define a connection to other cloud services and access them. A function could have zero or more bindings. There are input and output bindings, depending on whether you want to process data from a data service or want to output the processed result to another service – or both. An example would be reading messages from a queue and inserting the record into the document database. For more details, visit `https://learn.microsoft.com/en-us/azure/azure-functions/functions-triggers-bindings?tabs=csharp`.

While Azure Functions offers pay-as-you-go as the primary mode of billing, there are two more modes. The pay-as-you-go pricing plan is called the **consumption plan** while there is also a **premium plan** with higher performance and more options, such as network access. If you are already a user of Azure App Service (a generic hosting service for all sorts of web applications), you might already have an App Service plan. In this case, you can also add your functions to the same pricing plan. To learn more, visit `https://learn.microsoft.com/en-us/azure/azure-functions/functions-scale`.

Azure Functions supports running functions in both Linux and Windows. It supports a wide variety of languages, including C#, Java, JS, Python, PowerShell, TypeScript, and so on. It can also support other languages using a custom function handler. With Azure Functions, you also need to be aware of the runtime hosts. The latest version, 4, is GA. The previous versions have been mostly EOLed as of December 2022. More information on supported runtimes can be found at `https://learn.microsoft.com/en-us/azure/azure-functions/supported-languages`.

Now let us look at how functions are organized under a function app in Azure and how to create a function.

Function apps and creating your first function

There is the concept of a function app. It provides an execution context in which functions will run. Consider it as the higher-level abstraction on top of functions for managing and deploying one or more related functions as a unit. All functions inside a function app can be configured, deployed, and scaled together. They also share the same OS, runtime, billing, and deployment method.

When developing functions, a function app is represented by a project folder. This folder contains a host configuration file (`host.json`) for settings that are shared by all the functions, such as runtime configurations. The folder structure will change slightly based on the runtime you pick. For Python functions and the best practices to develop them, refer to the Python function developer guide: `https://learn.microsoft.com/en-us/azure/azure-functions/functions-reference-python`.

A function has two important parts, the first being the function code written in your runtime of choice and the configurations associated with it. These configurations are contained in a file called `function.json`. This file contains settings that specify the function trigger, bindings, and general function configuration. This is the V1 programming model of the function that is currently active. There is a V2 programming model in preview currently that uses decorators instead of configuring settings in `function.json`.

Now that we have a basic understanding of functions, let's go ahead and create a function app. We will do most of our workflow from the command line using Cloud Shell. We will be using the Python 3.9 runtime for our functions. Please note that creating a function and updating its code are two different processes. We will be using different CLIs and CLI options to do this. A function app name is one of the required parameters for this command. Now, let us use the Azure CLI to create the function app:

```
safeer [ ~ ]$ az functionapp create --consumption-plan-location
southindia --runtime python --runtime-version 3.9 --functions-version
4  --os-type linux --storage-account serverless101storage1 --name
serverless101firstAzFuncApp
Resource provider 'Microsoft.Web' used by this operation is not
registered. We are registering for you.
Registration succeeded.
Your Linux function app 'serverless101firstAzFuncApp', that uses a
consumption plan has been successfully created but is not active until
content is published using Azure Portal or the Functions Core Tools.
Resource provider 'microsoft.insights' used by this operation is not
registered. We are registering for you.
Registration succeeded.

{
  OUTPUT_TRUNCATED_AT_MULTIPLE_PLACES
  "defaultHostName": "serverless101firstazfuncapp.azurewebsites.net",
  "enabled": true,
  "enabledHostNames": [
    "serverless101firstazfuncapp.azurewebsites.net",
    "serverless101firstazfuncapp.scm.azurewebsites.net"
  ],
}
```

As you can see, the command throws a few messages and then creates the function app. The location for the consumption plan has to be provided as it may not be available in all regions. Given that we are choosing the Python runtime, it is important to also provide Linux as the OS since Python Azure functions can only run in Linux.

The command will also throw a whole bunch of output; I have truncated the result to only show the hostname endpoint of the function app. You might also see that it uses two Azure services called Azure Web (for hosting) and Insights (for monitoring). Although the function has been created, it won't be ready until we deploy some code to it, as pointed out by the first message in the command output.

Developing and deploying a Python function

Now that we have a function app, let us create a function inside it. You can have a single function or more than one related function inside a function app. All functions in a function app have to share the same runtime/language. To develop from the command line, the Azure Functions core tools need to be installed. On Cloud Shell, these will be available by default. If you want to set them up on your developer machine, read the instructions at https://learn.microsoft.com/en-us/azure/azure-functions/functions-run-local. This will allow you to also develop and deploy your functions locally for testing. It is also assumed that you have the correct Python runtime available on your local developer machine – in this case, Python 3.9. In order to start developing with Python, we need to first create a virtual environment as follows:

```
safeer [ ~/funcdev ]$ python3.9 -m venv .venv
safeer [ ~/funcdev ]$ source .venv/bin/activate
(.venv) safeer [ ~/funcdev ]$
```

When developing a function app with core tools, the first thing to do is create a project. Core tools provide a CLI named func. Let us use it to create the project folder:

```
(.venv) safeer [ ~/funcdev ]$ func init funcproject --worker-runtime
python
Found Python version 3.9.14 (python3).
Writing requirements.txt
Writing .funcignore
Writing getting_started.md
Writing .gitignore
Writing host.json
Writing local.settings.json
Writing /home/safeer/funcdev/funcproject/.vscode/extensions.json
```

As you can see, the command creates a number of files, including host.json, which we covered earlier. The getting_started.md file contains an explanation of each file and how to start from here. All the remaining commands should be run from within this folder. Also, note that we are using the V1 programming model for Python functions as V2 is still in preview. All actions from here onward have to be done after switching to the root of the project:

```
(.venv) safeer [ ~/funcdev ]$ cd funcproject/
```

Now, let us create a function. This requires a template and function name. Running func new without any arguments will list the available templates. Once the template and name are given, a directory with the name of the function will be created inside the project directory. It will contain boilerplate code that matches the function template chosen and the runtime of the project (__init__.py for Python). It will also contain a configuration file named function.json, which we covered briefly earlier.

Let us create the function:

```
(.venv) safeer [ ~/funcdev/funcproject ]$ func new
Select a number for template:
1. Azure Blob Storage trigger
2. Azure Cosmos DB trigger
3. Durable Functions activity
4. Durable Functions entity
5. Durable Functions HTTP starter
6. Durable Functions orchestrator
7. Azure Event Grid trigger
8. Azure Event Hub trigger
9. HTTP trigger
10. Kafka output
11. Kafka trigger
12. Azure Queue Storage trigger
13. RabbitMQ trigger
14. Azure Service Bus Queue trigger
15. Azure Service Bus Topic trigger
16. Timer trigger
Choose option: 9
HTTP trigger
Function name: [HttpTrigger] myHttpTriggerFunc
Writing /home/safeer/funcdev/funcproject/myHttpTriggerFunc/__init__.py
Writing /home/safeer/funcdev/funcproject/myHttpTriggerFunc/function.
json
The function "myHttpTriggerFunc" was created successfully from the
"HTTP trigger" template.
```

You can also get the list of supported templates using the following command:

```
func templates list -l python
```

First, let us look at the function.json file as it contains the bindings and triggers required by the function:

```
(.venv) safeer [ ~/funcdev/funcproject ]$ cat myHttpTriggerFunc/
function.json
{
  "scriptFile": "__init__.py",
  "bindings": [
    {
      "authLevel": "function",
      "type": "httpTrigger",
      "direction": "in",
```

```
      "name": "req",
      "methods": [
        "get",
        "post"
      ]
    },
    {
      "type": "http",
      "direction": "out",
      "name": "$return"
    }
  ]
}
```

There are a few points to note. They are as follows:

- `scriptFile`: This points to the file that defines the function.

- `bindings`: This defines all bindings, triggers, and outputs. This is an array of multiple bindings, and bindings have directions.

- The first binding defines the trigger, and the `name` key in the binding is the name of the input that will be passed to your function. In the case of an HTTP trigger, this will be an HTTP request.

- The second binding essentially defines the output of the function. In this case, it will return the return value of the function as an HTTP response.

Let us look at the boilerplate code inside `__initi__.py`:

```
(.venv) safeer [ ~/funcdev/funcproject ]$ cat myHttpTriggerFunc/__
init__.py
import logging
import azure.functions as func

def main(req: func.HttpRequest) -> func.HttpResponse:
    logging.info('Python HTTP trigger function processed a request.')

    name = req.params.get('name')
    if not name:
        try:
            req_body = req.get_json()
        except ValueError:
            pass
        else:
            name = req_body.get('name')
```

```
    if name:
        return func.HttpResponse(f"Hello, {name}. This HTTP triggered
function executed successfully.")
    else:
        return func.HttpResponse(
            "This HTTP triggered function executed successfully. Pass
a name in the query string or in the request body for a personalized
response.",
            status_code=200
        )
```

As you can see, the function handler – the `main()` function – is accepting a `req` argument of the `HttpResponse` type provided by the Azure Functions library. It also returns an object of the `HttpResponse` type from the same library. The rest of the code is self-explanatory. In short, the function accepts a `name` parameter either as a URL query parameter or as part of the JSON body. If you refer to the previous `functions.json` file, you will see that the input binding supports both `GET` and `POST` HTTP methods.

Now, let us run this function locally and test it:

```
(.venv) safeer [ ~/funcdev/funcproject ]$ func start --debug
```

This will throw a lot of debug info that you can look through, but the relevant lines will be the following:

```
Functions:

        myHttpTriggerFunc: [GET,POST] http://localhost:7071/api/
myHttpTriggerFunc
```

As you can see, the function is available at a local endpoint for us to test. Let us try this out:

```
safeer [ ~ ]$ curl http://localhost:7071/api/
myHttpTriggerFunc?name=Safeer
Hello, Safeer. This HTTP triggered function executed successfully.
```

The function worked successfully. Now, let us deploy this to the function app that we defined earlier:

```
safeer [ ~/funcdev/funcproject ]$ func azure functionapp publish
serverless101firstAzFuncApp
.....OUTUT_TRUNCATED.....
Functions in serverless101firstAzFuncApp:
    myHttpTriggerFunc - [httpTrigger]
        Invoke url: https://serverless101firstazfuncapp.
azurewebsites.net/api/myhttptriggerfunc?code=aFwqawKvnsG7LpPdqR_
jwZSuVU0IUU5GXeKZKmHlo6K2AzFuS7eD2g==
```

Let's hit the URL and pass the `name` argument to make sure it is working as expected:

```
safeer.cm$ curl "https://serverless101firstazfuncapp.
azurewebsites.net/api/myhttptriggerfunc?code=aFwqawKvnsG7LpPdqR_
jwZSuVU0IUU5GXeKZKmHlo6K2AzFuS7eD2g==&name=Azure"
Hello, Azure. This HTTP triggered function executed successfully
```

Note that I am running this from my laptop and not Cloud Shell, to make sure it is publicly accessible.

In real life, you will need to add your own business logic and add additional dependency libraries. The `requirements.txt` file in the project directory can be used for this. All dependencies will be installed at the function app level and all functions under an app are treated as one deployment unit. Refer to the developer guide to learn more about packaging, building, and deployment: `https://learn.microsoft.com/en-us/azure/azure-functions/functions-reference-python`.

In the next section, we will look into Azure Blob Storage.

Azure Blob Storage

Azure Blob Storage is an object storage offering from the Azure cloud. It is tuned for storing large amounts of data, in terms of the number of objects as well as the size of objects. **Blob** stands for **binary large objects** and images and videos are some of the file types that fall into this category. You can store any type and amount of unstructured data in blob storage. Some of the common usage patterns for blob storage are as follows:

- Media storage for images and videos to be centrally distributed
- Serving images or documents to the end customers directly
- Streaming multimedia content
- Data backup, archiving, and disaster recovery
- Data storage for big data analysis – known as data lakes

Like most other services, Blob Storage can be accessed from its REST API, the portal, the Azure CLI, or client SDKs in various languages. With SDKs, you can programmatically integrate Blob Storage into your applications.

Azure Blob Storage is created under a storage account. We briefly learned about storage accounts and used them when we covered functions. A storage account provides you with a unique namespace for all your storage needs, covering blobs, tables, queues, and so on. Under a storage account, blob storage is organized as containers.

Containers are how you organize your blobs. Consider them like folders in the traditional filesystem. In the case of AWS, these are Simple Storage Service buckets. You can attach policies and access control at the container level, depending on the type and purpose of blobs stored in that container. The policies applied at the container level are percolated to all blobs under it. A storage account can contain any number of containers, and containers can contain an unlimited number of blobs. Like the storage account, containers also need a unique name that is compliant with DNS names. Though we equated containers to a folder, they don't have any hierarchical structure like filesystems. All blobs stay under the same container at the same level. But to get around this, blob names can be prefixed with unique prefixes instead of folder names.

There are three types of blobs that Azure supports:

- **Block blobs**: Text and binary data that can be constructed and managed as individual blocks of data. Block blobs can go up to 190 TB.

- **Append blobs**: Similar to block blobs, but optimized for appending new data. These are good candidates for logging use cases.

- **Page blobs**: Store random access files of up to 8 TB in size. These are used for storing the virtual hard drive files of Azure virtual machines.

To learn more about the different types of blobs, refer to `https://learn.microsoft.com/en-us/rest/api/storageservices/understanding-block-blobs--append-blobs--and-page-blobs`.

Blobs also have their own naming convention and names can be as long as 1,024 characters. They are case-sensitive, have no restriction on the character combination (as long as special ones are escaped), and can contain path notations with `/`.

Now, let us see how to create and manage blob storage.

Creating and managing blob storage

Let us play around with Blob Storage a little bit now. We already have a resource group and storage account created, and the blob storage will be created under those. Before starting to use blob storage, you need to assign yourself permission to manage blobs by assigning yourself the Storage Blob Data Contributor role (roles are part of the access control system in Azure, called **Identity and Access Management (IAM)**). We will need our subscription ID (which you can get with `az account show`) and the resource group and storage account we created earlier. Replace your details accordingly:

```
safeer [ ~ ]$ az ad signed-in-user show --query id -o tsv
| az role assignment create      --role "Storage Blob Data
Contributor"      --assignee @-      --scope "/subscriptions/7u6f9085-
h345-4bnh-aa54-b030af67871e/resourceGroups/serverless101-rg/providers/
Microsoft.Storage/storageAccounts/serverless101storage1"
```

There will be some output about the role, scope, and so on. You can safely ignore it as long as you don't see an error. Now, let us go ahead and create a container in our storage account:

```
safeer [ ~ ]$ az storage container create --account-name
serverless101storage1  --name serverless101storage1container1  --auth-
mode login
{
  "created": true
}
```

Now, let us create a test file and upload it to the blob storage:

```
safeer [ ~ ]$ echo "Test blob" > firstBlob.txt

safeer [ ~ ]$ az storage blob upload --account-
name serverless101storage1 --container-name
serverless101storage1container1  --name firstBlob.txt --file
firstBlob.txt --auth-mode login
Finished[################################
##########################]  100.0000%
{
OUTPUT_TRUNCATED
}
```

We successfully uploaded our first blob. If a blob with the same name already existed, it will be overwritten. We will add one more blob, but this time, we will use a naming convention to represent a directory/file path:

```
safeer [ ~ ]$ echo "Test blob 2" > secondBlob.txt
safeer [ ~ ]$
safeer [ ~ ]$ az storage blob upload -account-
name serverless101storage1 -container-name
serverless101storage1container1  --name "firstDir/secondBlob.txt" -
file secondBlob.txt -auth-mode login
Finished[###################################
######################]  100.0000%
```

Now, let us list the files in the container:

```
safeer [ ~ ]$ az storage blob list --account-
name serverless101storage1 --container-name
serverless101storage1container1  --output table  --auth-mode login
Name                          Blob Type    Blob Tier    Length    Content
Type    Last Modified                 Snapshot
--------------------    -----------    -----------    --------    ---------
-----    ------------------------    ----------
firstBlob.txt                 BlockBlob    Hot          10        text/
plain         2023-03-08T12:04:01+00:00
```

```
firstDir/secondBlob.txt  BlockBlob     Hot           12          text/
plain        2023-03-08T12:10:34+00:00
```

As you can see, the blobs can have / in their name if you want your blobs to be organized like a filesystem. As a final step, let us download one of the files we uploaded:

```
safeer [ ~ ]$ az storage blob download --account-name
serverless101storage1 --container-name serverless101storage1container1
--name firstBlob.txt --file /tmp/blobFromContainer --auth-mode login
Finished[###############################
#########################]  100.0000%

safeer [ ~ ]$ cat /tmp/blobFromContainer
Test blob
```

This completes our practical section on Blob Storage.

Blob Storage and Azure Functions

As discussed in the *AzureFunctions* section, Azure functions interact with other cloud services using bindings. There are three types of triggers and bindings that a function can bind with Blob Storage:

- Run a function in response to a change in a blob. This will be a trigger.

- Read blobs from a function. This will be input binding.

- Write blobs to a container from a function. This will be output binding.

The following is a sample function.json file taken from the Azure documentation that covers a blob trigger:

```
{
    "scriptFile": "__init__.py",
    "disabled": false,
    "bindings": [
        {
            "name": "blobfile",
            "type": "blobTrigger",
            "direction": "in",
            "path": "serverless101storage1container1/firstDir/{name}",
            "connection":"MyStorageAccountAppSetting"
        }
    ]
}
```

The key elements in the file are explained as follows:

- The `name` key is the blob name that is passed to the function.

- The `path` key defines the blob pattern – the first part of the path is the container name while the rest of the part is the blob name. In this case, if a blob has the name `firstDir/secondBlob.txt`, then the `name` key will have the value `secondBlob.txt`.

- The `connection` key contains the connection string to the storage account where the blob container is located. This will be configured at the app level.

Here is some sample Python code from the Azure documentation for processing this trigger event:

```
import azure.functions as func
Import logging

def main(blobFile: func.InputStream):
    logging.info('Received the blob %s of %d bytes', blobFile.
name,blobFile.length)
```

The blob is provided to the function as an object of the Azure Functions `inputStream` class. For more information on this, refer to `https://learn.microsoft.com/en-us/python/api/azure-functions/azure.functions.blob.inputstream`.

The blob object has three attributes:

- The name of the blob

- The length of the blob in bytes

- A URI – the primary URL of the blob

As you can see, the code is sending the message to log as there is no output binding configured. If we expect to do further processing of the image and some information needs to be updated to another cloud service, an output binding will have to be defined.

This concludes our section on Blob Storage. Next, we will look into Cosmos DB.

Azure Cosmos DB

Cosmos DB is a fully managed, globally replicated, database platform offering from Azure. While it is technically a NoSQL platform, it is also a multimodel database, which means it provides more than one database API/protocol over this database. These APIs could be based on **relational data models (RDBMS)** or NoSQL models. The protocols supported by Cosmos DB and the corresponding Azure Cosmos DB offerings are as follows:

Cosmos DB for NoSQL	**Supports a native Azure NoSQL database**
Cosmos DB for MongoDB	MongoDB
Cosmos DB for Apache Cassandra	Apache Cassandra
Cosmos DB for Apache Gremlin	Graph database supporting the Gremlin graph traversal language based on the Apache TinkerPop specification
Cosmos DB for PostgreSQL	PostgreSQL
Cosmos DB for Tables	Supports Azure Table storage, which can be used to store large amounts of structured NoSQL data as a key-attribute store with a schemaless design

Table 4.1 – Cosmos DB supported protocols

As you can see, Cosmos DB has taken multimodel to the next level with support for such a diverse set of relational and NoSQL models. Let us look at some of the key benefits of Cosmos DB:

- Real-time data writes and reads with minimal latencies across the globe.
- It supports multi-region writes and data replication to any chosen Azure region.
- Encryption at rest and transit with self-managed keys.
- Autoscaling to cope with steady and sporadic traffic growth.
- A wide range of client SDKs.
- Change feed for change data capture.
- Interactions with functions as triggers and bindings.
- Schema-agnostic indexing.
- Auto-backups without impacting performance.

Now that we understand the benefits of Cosmos DB, let us look at the key elements of it.

Elements of Cosmos DB

The top element in Cosmos DB is the Cosmos DB account, similar to a storage account, which could hold different types of data. Within the account, you can create one or more databases. Databases are like virtual namespaces for your data for a specific purpose. Based on the Cosmos DB API you use, the database will represent different things. For example, for MongoDB/NoSQL and Gremlin, this is called the database itself, while for Apache Cassandra, this will be a keyspace. Within the database, you can create one or more containers to store and manage your data. Depending on the API used, the container could be a relational database table or a graph in Gremlin, or a collection in MongoDB or Cassandra. A container can contain a number of *items*, which are individual units of data. Similar to the other concepts, based on the APIs used, an item could be a NoSQL document, an RDBMS row, a graph node/edge, and so on.

Cosmos DB offers two pricing models based on the way in which the capacity is provisioned and scaled: the provisioned throughput model and the serverless model. In the provisioned model, you can either set a manual (standard mode) throughput limit or let it autoscale. These settings can be applied at the database or container levels. With serverless Cosmos DB, you can manage your capacity in a consumption-based manner (similar to how we set the plan for Azure Functions earlier). Cosmos DB has the concept of **Request Units** (**RUs**), which can be used to express any database operations in a standard pattern. In provisioned capacity mode, you have to set the number of RUs per second as the capacity. In serverless mode, you don't have to set this limit and Cosmos DB will scale up or down automatically.

Data partitioning and partition keys

We learned about how databases and containers are used to organize Cosmos DB databases. Beyond these two, Cosmos DB uses partitioning to scale containers. Partitioning is the method of dividing the items in a container into distinct subgroups. The top-level logical grouping is called logical partitions. For this, a partition key is associated with every item in the container, and the logical partition is formed with all the items with the same partition key.

In addition to being associated with a partition key, an item also has an item ID that uniquely identifies that item within the logical partition. The index or primary key (in RDBMS language) that identifies an item uniquely is created by combining the partition key and the item ID. Every logical partition has an upper limit of 20 GB. So, it is important to choose and model your partition strategy carefully to pick the right partition key.

Within a container, physical scalability is achieved by the use of physical partitions. One or more logical partitions are mapped to a physical partition. This is a Cosmos DB construct that is managed internally and the division and distribution of physical partitions depend on the requests they will serve and the size of the data the underlying logical partitions hold. A partition has an upper limit of serving 10,000 RUs and storing 50 GB of data. As your data size and traffic requirements grow, Azure will automatically split existing physical partitions to match the growth. Unlike logical partitions, you do not have control to model the physical partitions using the partition key or any other parameters.

Replica sets are how Cosmos DB makes the databases durable and globally distributed. Replica sets, as their name suggests, keep copies of the data that the physical partition holds and each copy is called a replica. All replicas together serve the traffic for a physical partition. To learn more details about partitioning and partition strategies, visit `https://learn.microsoft.com/en-us/azure/cosmos-db/partitioning-overview`.

Creating and managing Cosmos DB

Now that you have learned about the breadth and depth of the Cosmos DB portfolio, let us bring our focus to one part of Cosmos DB that is most interesting to us – serverless Cosmos DB. As we discussed already, this is just a different mode of operation for Cosmos DB and the serverless mode is applicable to most of the supported APIs. We will start by creating a Cosmos DB account, followed by creating a serverless database, and then we'll see how to access it.

Let us first create a Cosmos DB account:

```
safeer [ ~ ]$ az cosmosdb create --name serverless101cosmos
--resource-group serverless101-rg  --default-consistency-level
Eventual --capabilities EnableServerless
Resource provider 'Microsoft.DocumentDB' used by this operation is not
registered. We are registering for you.
Registration succeeded.
{
 …..OUTPUT TRUNCATED……
   "capabilities": [
     {
       "name": "EnableServerless"
     }
 …..OUTPUT TRUNCATED……
   "documentEndpoint": "https://serverless101cosmos.documents.azure.
com:443/",
 …..OUTPUT TRUNCATED……
}
```

As you can see, we created a serverless Cosmos DB account with the command-line option `--capabilities EnableServerless`. Now, let us create a database and container inside this account for NoSQL:

```
safeer [ ~ ]$ az cosmosdb sql database create --account-name
serverless101cosmos --resource-group serverless101-rg --name
serverlessdb101


safeer [ ~ ]$ az cosmosdb sql container create --account-name
serverless101cosmos --resource-group serverless101-rg --database-name
serverlessdb101 --name serverlessdb101container1 --partition-key-path
```

```
/documents
{
...OUTPUT_TRUNCATED....
  "name": "serverlessdb101container1",
...OUTPUT_TRUNCATED....
    "partitionKey": {
      "kind": "Hash",
      "paths": [
        "/documents"
      ],
...OUTPUT_TRUNCATED....
  "tags": null,
  "type": "Microsoft.DocumentDB/databaseAccounts/sqlDatabases/
containers"
}
```

Now that the container is created, you can use any of the Cosmos DB SDKs to perform operations on the items in the container. For examples from the Python SDK, refer to https://learn.microsoft.com/en-us/azure/cosmos-db/nosql/samples-python.

Cosmos DB and Azure Functions

Azure Cosmos DB provides three bindings for Azure Functions – triggers, input binding, and output binding:

- **Triggers** – Functions invoked when a Cosmos DB document is created or updated
- **Input Bindings** – Read a Cosmos DB document
- **Output Bindings** – Modify a Cosmos DB document

To learn more about the bindings, refer to https://learn.microsoft.com/en-us/azure/azure-functions/functions-bindings-cosmosdb-v2 and https://learn.microsoft.com/en-us/azure/cosmos-db/nosql/serverless-computing-database.

Up next, we will quickly explore the event and messaging services.

Azure event and messaging services

Azure has a number of similar-sounding event and messaging services with overlapping functions. We will take a quick look at all of them.

Azure Event Grid

Event Grid is suited for event reaction systems, where you want to programmatically respond to events that occur in another cloud service. It follows a publisher-subscriber model. It has native integration with a large number of Azure services as well as third-party services. It routes events efficiently and reliably from the event providers to the registered subscriber endpoints. As this is an event routing system, the event in itself contains only the information about the event and not the actual data. For example, Event Grid can tell you when a new blob is created in a storage container along with the metadata about the blob, but it can't provide the actual content of the blob.

Event Grid comes in two flavors – one is the default Azure Event Grid, which provides the **Platform as a Service (PaaS)** events, and the other one is Event Grid on Kubernetes with Event Arc, which allows you to deploy Event Grid on any Kubernetes cluster, inside or outside Azure. You can use event filters to route specific events to different endpoints.

Azure Event Hubs

Event Hubs is a big data platform that can ingest a large number of events in a short time. Data can be received from a number of concurrent sources, then retained and replayed if necessary. Like Event Grid, it can also integrate with many Azure services, such as Stream Analytics, Event Grid, Power BI, and so on, along with non-cloud services such as Apache Spark. In essence, it is a distributed streaming platform that is tuned for high throughput and performance. It supports both real-time and batch processing of its stream data.

Event Hubs exposes the **Advanced Message Queuing Protocol (AMQP)** 1.0 supported endpoint for all communications with its clients (producers and consumers). As a time-tested industry standard for messaging, AMQP has a rich ecosystem of SDKs and clients that allow integration in many ways. In addition, Event Hubs also allows Kafka clients to talk to it using the Event Hubs for Apache Kafka ecosystem. Event Hubs is suited for streaming data such as logs, IoT data, metrics, and so on.

Azure Service Bus

This is a fully managed message broker that supports both message queues and topics. It is used to decouple enterprise applications from each other and provide a communication bus. It allows encapsulating and sending messages (with data and metadata) between applications using common data encoding formats such as JSON, XML, Avro, and so on. Like other messaging systems, Service Bus also decouples the producers and consumers of the message, and one doesn't have to be necessarily online when the other is. It also improves consumer load balancing, with competing consumers getting exclusive access to portions of the data.

An important feature of Service Bus is that it supports executing multiple operations as a single transaction. It has support for features such as dead-lettering, auto-forwarding, scheduled delivery, and so on. Subscribers can also set rules to filter the messages and receive only what they need.

Azure Event Grid with Azure Functions

Now that we know about the various event and messaging services, let us zoom in on the service that is of most interest to us – **Event Grid**. With Event Grid, you can build a scalable event response system. It connects data sources and event handlers using Event Grid topics and subscriptions. The following table lists the core concepts of Event Grid:

Events	State change or event occurrence notification
Event Source	The services or systems that are generating the events
Topics	The endpoint to which event sources publish their events
System Topics	In-built topics provided by other Azure services
Event Subscription	The configuration and endpoint to filter and route messages to the event handlers based on defined criteria
Event Handler	The application or cloud service that consumes the events

Table 4.2 – Event Grid core concepts

To learn more about these concepts, refer to `https://learn.microsoft.com/en-us/azure/event-grid/concepts`.

Event Grid has three bindings for Azure Functions:

- An event published to the Event Grid topic can trigger a function.
- An Azure function can publish an Event Grid custom topic.
- You can also have an HTTP trigger if you want to return a custom error code to Event Grid for a retry of event delivery.

To learn more about the bindings, visit `https://learn.microsoft.com/en-us/azure/event-grid/receive-events`.

Now that we have an understanding of Event Grid, let us create a topic and subscription:

```
safeer [ ~ ]$ az eventgrid topic create --resource-group
serverless101-rg --name serverless101topic1
Resource provider 'Microsoft.EventGrid' used by this operation is not
registered. We are registering for you.
Registration succeeded.
{
... OUTPUT_TRUNCATED...
  "endpoint": "https://serverless101topic1.southindia-1.eventgrid.
azure.net/api/events",
```

```
... OUTPUT_TRUNCATED...
}
```

We will create a subscription to this topic with the function we created in the earlier section as the function handler.

Let us first retrieve the IDs of our topic and function as they are to be passed as arguments to the event subscription creation command:

```
safeer [ ~ ]$ az functionapp function list -g serverless101-rg -n
serverless101firstAzFuncApp --query '[0].id'
"/subscriptions/7u6f9085-h345-4bnh-aa54-b030af67871e/
resourceGroups/serverless101-rg/providers/Microsoft.Web/sites/
serverless101firstAzFuncApp/functions/myHttpTriggerFunc"

safeer [ ~ ]$ az eventgrid topic list -g serverless101-rg --query
'[0].id'
"/subscriptions/7u6f9085-h345-4bnh-aa54-b030af67871e/resourceGroups/
serverless101-rg/providers/Microsoft.EventGrid/topics/
serverless101topic1"
```

Now, we will create a subscription on the topic with the `myHttpTriggerFunc` function as the event handler. Note that this is just provided as an example and this particular command will fail since the function was not configured for binding with Event Grid, as you can see in the error message:

```
safeer [ ~ ]$ az eventgrid event-subscription create
--name serverless101subscription1 --source-resource-id /
subscriptions/7u6f9085-h345-4bnh-aa54-b030af67871e/resourceGroups/
serverless101-rg/providers/Microsoft.EventGrid/topics/
serverless101topic1 --endpoint /subscriptions/7u6f9085-h345-4bnh-
aa54-b030af67871e/resourceGroups/serverless101-rg/providers/Microsoft.
Web/sites/serverless101firstAzFuncApp/functions/myHttpTriggerFunc
--endpoint-type azurefunction
(Unsupported Azure Function Trigger) Can't add resource /
subscriptions/7u6f9085-h345-4bnh-aa54-b030af67871e/
resourceGroups/serverless101-rg/providers/Microsoft.Web/sites/
serverless101firstAzFuncApp/functions/myHttpTriggerFunc as a
destination with unsupported Azure function triggers. Azure Event Grid
supports EventGrid Trigger type only.
```

You can find an example with the HTTP endpoint as the handler at this link: `https://learn.microsoft.com/en-us/azure/event-grid/scripts/event-grid-cli-subscribe-custom-topic`.

Now that you have got a reasonable idea about events and serverless in Azure, let us move on to the last service in the list, Azure Logic Apps.

Azure Logic Apps

Azure Logic Apps is a workflow orchestration PaaS offering from Azure. It provides a simplified and intuitive user interface and user experience to design, develop, and orchestrate workflows. You can integrate these workflows with a large number of Azure cloud services as well as other vendors' services. This allows you to model business processes and automate them, thus offering a powerful business process automation service. Since this is a PaaS, you can scale your workflows to the scale of your business needs.

Some of the key points to remember about the workflow design are as follows:

- A workflow is started by a trigger – this could be a webhook, an email, an Azure cloud event, and so on.

- The trigger kicks off a sequence of one or more *actions* that carry out an operation that is part of the workflow.

- The workflow also provides control structures for conditional logic and loops.

- At each action (aka step of the workflow), you can also do data transformations to filter or convert the data you received from a previous action or the data to be sent to the next action.

- There are prebuilt actions and triggers that integrate with Azure and Office 365 services, as well as third parties. These ready-to-use actions are called connectors and are managed by Microsoft.

- If none of the prebuilt connectors suit your needs, you can create your own connector, or use built-in HTTP operations or request triggers to talk to an external API.

Some of the common use cases for workflows are automated notifications for events that occur in the cloud, the multi-step processing of a blob uploaded to Blob Storage, examining APIs and taking action, and so on.

Key concepts of Logic Apps

Some of the key terms related to Logic Apps are as follows:

- **Logic app**: The Azure resource that hosts your workflows. There are two types of logic app resources – **Standard** and **Consumption**:

  - The consumption logic app can run in multi-tenant logic app environments as well as integration service environments.

  - The standard logic app can only run in single-tenant logic app environments.

  - Depending on the environment and resource type, there will be differences in some of the features supported, such as network/VPC access, resource sharing, and pricing plan.

- For a detailed comparison, refer to `https://learn.microsoft.com/en-us/azure/logic-apps/single-tenant-overview-compare`.

- **Workflow**: The logical representation of the process that needs to be automated, defined as a series of steps and started by a trigger:

  - With the consumption resource type, you can only have one workflow per service, whereas with the standard resource type, you can have multiple workflows.

  - With the standard resource type, you can create both stateful and stateless workflows.

  - A stateful workflow persists the data and details of previous runs to disk storage. This can be used to review the run history as well as to reconstruct a workflow run from the saved state.

- **Trigger**: The condition or event that should occur in order for the workflow to be initiated. This could be an email being received, a tweet, a file uploaded to Blob Storage, and so on.

- **Action**: Each step in the workflow after the trigger is called an action. An action represents an operation within the workflow.

- **Connector**: A connector provides an interface that allows you to work with the data, events, or resources in other Azure services, SaaS products, systems, platforms, and so on. They can also be used to control your workflow. Connectors offer either actions, triggers, or both. You can use them to perform one step of your workflow. Connectors will have configurable properties that will allow you to model your workflow and steps. There are two types of connectors:

  - `Builtin`: These are connectors that help you control your workflows. They allow you to schedule workflows, manipulate data, control steps (with conditionals, loops, and so on), run custom code (inline or using Azure Functions), and perform operations on a selected set of services (Azure and non-Azure) that offer built-in connectors.

  - `Managed`: These connectors are proxies or wrappers to a REST API that you can use to access apps, data, services, and so on. Unlike built-in connectors, you need to first create a connection from your workflow and authenticate your identity in order to start using the service. Note that the same service might provide both managed and built-in connectors. All connectors are managed by Microsoft. Based on the logic app resource type and environment you use, the list of connectors available to you will vary.

- **Integration Account**: An integration account resource is a scalable cloud-based container in Azure that helps you store and manage B2B artifacts that you define and use in your workflows for B2B scenarios. This is needed if you are building B2B and enterprise integration workflows.

Now that you know about the basic building blocks of Azure Logic Apps, let us look at creating a simple workflow using the Azure CLI.

Creating a Logic Apps workflow

The best way to create logic apps currently is to use the Azure portal. Given that we are experimenting with workflows, it is easier to explore things using the portal visually. When you create a logic app on the portal, as the last step, it opens up a workflow designer where you can drag and drop connectors and steps and configure them in a visual way. Demonstrating that in a book would need a lot of screenshots just for one workflow. So, for now, we will stick to the CLI.

Before we create a logic app using the CLI, we need to install the Azure Logic Apps CLI extension. This can be achieved by running this command:

```
az extension add --name logic
```

Extensions are how the Azure CLI extends its capabilities and adds new features. Logic Apps is not part of the CLI's core capability and is currently in preview. This means that logic app creation with the CLI is not covered under Azure customer support and it is not recommended to use in production environments. But given that this was released in mid-2022, we can expect that this will soon come out of preview and be certified for production use.

When creating a workflow via the CLI, we also need to have a way to define a workflow with its trigger and steps. The way the CLI does this is by taking a JSON file as an input, which has a defined schema as described here. We will use a simple workflow definition shown as follows. We will first use this schema to create the logic app and then explain the components of the definition. The workflow has to be saved in a file as follows:

```
safeer [ ~ ]$ cat workflow.json
{
  "definition": {
    "$schema": "https://schema.management.azure.com/providers/
Microsoft.Logic/schemas/2016-06-01/workflowdefinition.json#",
    "actions": {
      "Terminate": {
        "inputs": {
          "runStatus": "Succeeded"
        },
        "runAfter": {
          "myHttpTriggerFunc": [
            "Succeeded"
          ]
        },
        "type": "Terminate"
      },
      "myHttpTriggerFunc": {
        "inputs": {
          "function": {
```

```
            "id": "/subscriptions/7u6f9085-h345-4bnh-aa54-
b030af67871e/resourceGroups/serverless101-rg/providers/Microsoft.Web/
sites/serverless101firstAzFuncApp/functions/myHttpTriggerFunc"
        },
        "method": "GET",
        "queries": "@triggerOutputs()['queries']"
      },
      "runAfter": {},
      "type": "Function"
    }
  },
  "contentVersion": "1.0.0.0",
  "outputs": {},
  "parameters": {},
  "triggers": {
    "manual": {
      "inputs": {
        "method": "GET",
        "relativePath": "/startworkflow"
      },
      "kind": "Http",
      "type": "Request"
    }
  }
},
"parameters": {}
}
```

You can see the following:

- The trigger is an HTTP request that accepts a GET request at the /startworkflow path.

- When the workflow is created, it will return the endpoint that we need to hit in order to trigger this workflow.

- The workflow starts with the action named myHttpTriggerFunc, which runs an Azure function that we created in the *Azure Functions* section.

- We are passing the query parameters that the trigger receives into this function.

- The next step is a terminate step, which essentially returns a success and finishes the workflow.

- You can see that the actions are ordered using the runAfter keyword.

- The rest of the code is boilerplate code that is common to all workflow definitions.

Now let us create the workflow using the CLI:

```
safeer [ ~ ]$ az logic workflow create --resource-group
serverless101-rg --name firstLogicApp --definition workflow.json
{
   "accessControl": null,
   "accessEndpoint": "https://prod-11.southindia.logic.azure.com:443/
workflows/e68dae23c4c641fa95ebcefccd0a9a92",
 ...OUTPUT_TRUNCATED....
   "id": "/subscriptions/7u6f9085-h345-4bnh-aa54-b030af67871e/
resourceGroups/serverless101-rg/providers/Microsoft.Logic/workflows/
firstLogicApp",
...OUTPUT_TRUNCATED....
}
```

As you can see, the CLI provided accessEndpoint and the ID of the Logic App to us. Unfortunately, an issue with this is that the trigger URL needs an authorization code and API version along with the endpoint, which we can't retrieve from the CLI. So, you need to do the following:

1.   Go to https://portal.azure.com and search for Logic Apps.

2.   Click **Logic Apps**. The list will show firstLogicApp.

3.   Click the function.

4.   Click the **Edit** button. This will open the workflow designer view.

5.   Click on the trigger titled When an HTTP request is received. It will expand to a tile where you can find HTTP GET URL.

6.   Copy this URL. It will be something like the following:

     https://prod-11.southindia.logic.azure.com/workflows/
     e68dae23c4c641fa95ebcefccd0a9a92/triggers/manual/paths/invoke/
     startworkflow?api-version=2016-10-01&sp=%2Ftriggers%2Fmanual%
     2Frun&sv=1.0&sig=Db-GCsA_Gb6L0G0dTiDpp_XGloRksC3U0OkjqWyxLSU

As you can see, the base URL is accessEndpoint, with different parameters as the URL encoded. Remember that our function takes name=SomeName as the argument. We are going to append that to the URL and invoke it to trigger the workflow:

```
safeer [ ~ ]$ curl -v -s "https://prod-11.southindia.logic.
azure.com/workflows/e68dae23c4c641fa95ebcefccd0a9a92/triggers/
manual/paths/invoke/startworkflow?api-version=2016-10-01&sp=%
2Ftriggers%2Fmanual%2Frun&sv=1.0&sig=Db-GCsA_Gb6L0G0dTiDpp_
XGloRksC3U0OkjqWyxLSU&name=Safeer"
...OUTPUT TRUNCATED....
< HTTP/2 202
...OUTPUT TRUNCATED....
```

This command will return without any output, but if you see the debug info printed on the screen, it will show that an HTTP 202 is returned. This is because workflows run asynchronously and Azure has accepted your submission. Now, to see the output, you have to go back to the portal and navigate to the Logic Apps page as before. Once on the main page, there is a **Runs history** tab. Click that and you will see an entry for the time when you ran the `curl` command. It should show it as successful. Click on the entry, and this will take you to a window similar to the designer. But this time, if you click the tile for the `myHttpTriggerFunc` function, you will see the `OUTPUT` body showing the message `Hello, Safeer. This HTTP triggered function executed successfully`, as we would expect from the Azure function.

This concludes our section on workflows. Next, we will look into a small project that we can do using Azure serverless services.

Project – image resizing with Azure Functions

In this project, we will resize every image that is uploaded to Azure Blob Storage. The architecture diagram is as follows:

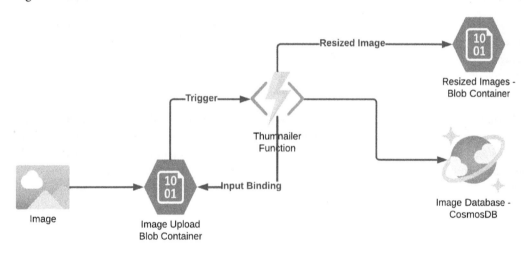

Figure 4.3 – Image resizing with Azure Functions

This is how the process will work:

- **Image Upload**: An image is uploaded to the Azure Blob Storage container.
- **Image Processor**: Each upload triggers an Azure function.
- **Resized Images**: The image processor will fetch the image file based on the trigger event, resize it to a standard size, and then upload it to another Blob Storage container.

- Then, the function will update the details of both the original image and the resized image in a Cosmos DB NoSQL database.

This project with its design, setup instructions, and code is available at the GitHub link: `https://github.com/PacktPublishing/Architecting-Cloud-Native-Serverless-Solutions/tree/main/chapter-4`.

This concludes our chapter on Azure serverless.

Summary

Microsoft frequently adds more services to its Azure portfolio. In addition to having serverless services that match AWS, they also have a large number of vendor integrations. These integrations really help in building enterprise serverless applications. There are more services in the serverless portfolio to cover, but that would merit a book on its own. For some of the customer success stories, visit `https://customers.microsoft.com/` and search for any of the Azure serverless services. There are a number of Azure Functions case studies.

In this chapter, we started by learning the fundamentals of Azure, followed by covering the fundamental serverless service – Azure Functions, which offers a FaaS platform. Then we went through Blob Storage, Cosmos DB, Event Grid, and finally, Logic Apps. This way, we were able to touch upon the core serverless services: FaaS, Object Storage, eventing, NoSQL databases, and workflow management. We concluded with a project on image resizing.

As I mentioned, there are a lot more services that we couldn't cover here, and also a number of features of the services we covered. For example, Durable Functions is an extension of Azure Functions that brings statefulness to functions. The topics covered in this chapter should provide you with a solid foundation and enable you to explore advanced services and concepts of Azure serverless. In the next chapter, we will cover Google Cloud and its serverless platforms.

5

Serverless Solutions in GCP

Google Cloud is one of the three top cloud vendors today. While AWS always had the upper hand in delivering new services and adding new regions, GCP is quickly catching up with its equivalent services. Like AWS powering the Amazon e-commerce business, most of the GCP services have been powering Google's product arrays, including GSuite, Search, YouTube, and others. This means that they have been battle tested for global workloads and are highly reliable.

It is interesting to note that GCP started with an early version of what could be called a serverless service. Google AppEngine was a trendsetter of its time in running application workloads without requiring any services to be provisioned. It was a shift from the tried and tested web hosting model and gave a lot of freedom to developers. When AWS became a formidable force in the cloud and GCP was not very fast to catch up in the initial years, the serverless trend was categorically set by AWS. But as GCP woke up from its slumber and started expanding its product portfolio, so did its serverless offerings.

In this chapter, we will start by covering the basics of GCP and how to get started with it. Then, we will cover some of the core serverless offerings from them. This will be followed by short coverage of a few more services and end with a practical project that uses a set of chosen serverless services:

- Prerequisites and GCP basics
- Google Cloud Functions
- GCP Pub/Sub
- Google Cloud Storage (GCS)
- Google Cloud Workflows
- The project – nameplate scanning and traffic notice

Prerequisites and GCP basics

The main prerequisites for this chapter are to have a GCP account and to set up your developer environment. For that, you need to create a Google account or use an existing one. You will use this account to sign up with Google Cloud. You will be given $300 worth of credit for free as a one-time offer from GCP. If you have already used this up, you will have to provide additional payment options. There is the concept of a billing account, where you must add your payment information and then connect it to your cloud accounts or projects. Google Cloud can also be managed via its web console or using the CLI, in addition to its APIs. We will mostly stick to using the command line for our work as it is easier to replicate and less prone to changes over time.

Before we set up an account, here are a few basics so that you can understand the concepts that will be discussed:

- Most of Google Cloud's resources are hosted in all or a large number of locations globally. There are two key terms when it comes to locations in GCP – regions and zones. A region is a geographic location where specific services are run and you can host your applications. Some services are region-specific – such as static IP addresses – where the resource you created is tied to that geography and the Google infrastructure of that region. Some other services are global and have no references to a region – while the underlying infrastructure might cover one or more regions, that is not the user's concern.

- Zones are like closely tied data centers within a region. A typical region has three or more zones. Many resources are zone-specific, such as virtual machines. Regional resources can be used by any other service from within that region, irrespective of the zone, while resources in the zone can only be used from within that zone. This is equivalent to AWS regions and availability zones. More information can be found here: `https://cloud.google.com/compute/docs/regions-zones`.

- GCP uses a resource hierarchy for customers to organize their cloud workloads according to their organizational structure. We will cover the most important points here. We will discuss this structure in a bottom-up way. At the lowest level is GCP resources, which can include things such as virtual machines, GCS buckets, Pub/Sub queues, IP addresses, and so on. These resources have exactly one parent, which is known as a project. Projects can be contained within folders, which can have parent and sibling folders. Folders are children of an organization; usually, one organization represents one customer or business unit. You can learn more about GCP's resource hierarchy at `https://cloud.google.com/resource-manager/docs/cloud-platform-resource-hierarchy`.

Now that we have a basic idea of how GCP is organized, let's sign up for GCP by heading to `https://cloud.google.com/gcp/`. As mentioned earlier, if you are a newcomer here, you will get $300 worth of credit to experiment with. Once you have signed up, you will get some default projects. Normally, to start working on the cloud, you will need to use their graphical console – which is where you are probably now if you just signed up for an account. The cloud console can be accessed at `https://console.cloud.google.com`. A large portion of what we intend to do can be done from this console, but we will favor the command line mostly.

In a typical developer workflow, to work from the CLI, you must get your cloud credentials, set them up on your developer machine/laptop, and install and use the available CLIs. GCP has made this easier with Cloud Shell. It is a web shell that you can access directly from your cloud console. You can find it at `https://shell.cloud.google.com` or click on the shell icon in the top-right corner of your cloud console. The good thing with Cloud Shell is that you don't have to explicitly set most of your credentials or install the default CLIs. These are all pre-provisioned for you. Then, when you need authorization and/or the existing authorization expires, the shell will launch a window, asking you to authorize. This makes authentication much safer and easier.

Another thing to note is that to use any GCP service, you need to first enable the corresponding APIs. You can do this from the console by visiting `https://console.developers.google.com/apis` and picking the APIs/services you want to enable. Another way is to use the CLI – called the gcloud CLI. We will be going the CLI route.

First, you need to create a project and then set it as your project for the current shell context:

```
safeercm@cloudshell:~$ gcloud projects create packt-serverless101
Create in progress for [https://cloudresourcemanager.googleapis.com/
v1/projects/packt-serverless101].
Waiting for [operations/cp.6982808765391966382] to finish...done.
Enabling service [cloudapis.googleapis.com] on project [packt-
serverless101]...
Operation "operations/acat.p2-579396612070-c472bb5a-24b1-4204-9559-
023a8df1db7b" finished successfully.

safeercm@cloudshell:~$ gcloud config set project packt-serverless101
Updated property [core/project].
safeercm@cloudshell:~ (packt-serverless101)$
```

You can also set the default region for your services in the config. For example, to set the default regions for Cloud Functions, run the following command:

```
safeercm@cloudshell:~ (packt-serverless101)$ gcloud config set
functions/region asia-south1
Updated property [functions/region].
```

This sets the default region where I will create functions to Mumbai, India.

Now that we have set up the basic developer environment, let's learn about our first serverless service: Cloud Functions.

Cloud Functions

Cloud Functions is the **Function-as-a-Service (FaaS)** offering from GCP. Premiered in 2017, it was envisioned to provide a simple and intuitive developer experience for deploying lightweight code snippets that responds to events. Similar to other vendors' FaaS offerings, it comes with the same baseline promises – pay for what you use, billed only for execution time and resource usage, scale up and down as and when required, and so on. What distinguishes the serverless compute offerings from GCP – Cloud Functions and its powerful cousin, Cloud Run – is that they are based on an open source FaaS framework called **Knative**. This helps in portability as you can run them in local environments, on-premises, with GCP, as well as any other platform that supports Knative, such as Red Hat OpenShift.

Core features

Cloud Functions are lightweight functions built for a single purpose. They provide all the usual features that most FaaS vendors provide:

- End-to-end software development life cycle management with debugging support.
- Autoscaling to handle global scale workloads.
- No infrastructure management.
- Event-driven architecture capabilities with support for a large number of event sources.
- You pay for what you use. The pricing strategy will be discussed later.
- You can avoid lock-in since Cloud Functions supports open source Knative-based serverless environments
- Connection to and extension of Google's other cloud services.

There are two variants of Cloud Functions: 1st generation and 2nd-generation. While they share many common characteristics, some changes affect both their costs and the technologies they use behind the scenes.

The pricing model

Function pricing is fundamentally modeled based on the number of times the function runs, how long it runs, and how many computing and networking resources it uses. There are a few additional caveats to this billing model, as follows:

- Outbound network requests – out of GCP – incur a data transfer fee.

- Function images are stored either in the container registry or artifact registry. The container registry is billed for usage, while the artifact registry (used by default in 2nd gen) will start costing you if you cross the free tier usage.

- The function images are built using the Cloud Build service and have their own pricing.

- 2nd gen uses the Eventarc service for event management and is billable.

- Various functions are associated with Firebase's services and have different pricing.

- Compute is the fundamental unit for cost calculation and depends on two variables that are measured over time – the memory it uses (measured in GB-seconds) and the CPU used (measured in GHz-seconds).

- The function minimum concurrency – how many invocations you run in parallel at a minimum – also impacts billing in terms of idle time.

- Outbound data and disk usage are measured in GBs.

Don't be alarmed by these details if you are playing with functions and learning; there is a free tier that is ample enough for your usage and you should not worry about it. But if you are a cloud architect modeling your service around serverless, you need to go into the details. More up-to-date information and sample calculations can be found at `https://cloud.google.com/functions/pricing`

Operating system (OS) and runtime support

Similar to AWS Lambda, Cloud Functions provides standard **OS** images to run the function code. Cloud Functions supports several programming languages. At the time of writing, the supported OSes are Ubuntu 22.04 and Ubuntu 18.04, both of which are the **long-term support** (**LTS**) variants of Ubuntu.

Several programming languages are also supported by Cloud Functions. To run these functions in the supported languages, a base container with a supported OS and language runtime needs to be provided. For Cloud Functions, these choices are already made for you by Google. At the time of writing, the following programming language version and OS version combos are supported:

	Ubuntu 18.04	Ubuntu 22.04
Node.js	6, 8, 10, 12, 14, 16	18
Python	3.7, 3.8, 3.9	3.10, 3.11
Go	1.11, 1.12, 1.13, 1.16	1.18, 1.19
Java	11	17
Ruby	2.6, 2.7, 3.0	No support
PHP	7.4, 8.1	No support
.NET Core	3	6

Table 5.1 – Cloud Functions OS and runtime support

Some of the additional features of Cloud Functions are as follows:

- All language dependencies are resolved at runtime
- You can assign deploy time environment variables for customization
- You can deploy to one or more regions at the same time
- Cloud Functions supports large amounts of memory allocation (large enough for a function)
- Privilege restrictions and security are provided via per-function identities

Now that we have a grasp of the basic features and pricing of Cloud Functions, let's jump right into experimenting with this service.

Function triggers

Functions are usually run in response to something that has happened. These are called triggers and are associated with the function. There are two broader categories of triggers that Cloud Functions supports:

- **HTTP triggers**: Functions are run in response to an HTTP request and are called HTTP functions
- **Event triggers**: Functions are invoked in response to an event that occurred within your GCP system and are called event-driven functions

A function can only have one trigger, while an event can trigger multiple functions. You must associate any function with exactly one trigger.

Within event triggers, there are three important categories:

- Pub/Sub triggers
- Cloud storage triggers
- Eventarc triggers

Since Pub/Sub and cloud storage are foundational to the serverless ecosystem, we will look into these triggers in the corresponding sections for these services. But for now, we will experiment with functions that are triggered by HTTP.

Function's structure and dependency management

An important characteristic of any function is its **entry point**. GCP defines function entry points as follows:

> *"Your source code must define an entry point for your function, which is the particular code that is executed when the Cloud Function is invoked. You specify this entry point when you deploy your function."*

Since the entry point handles the life cycle of the function, the input to the function is passed as arguments to it. The type of arguments will differ slightly based on what kind of trigger is meant to be handled by it. Also, keep in mind that for certain languages, entry points will be classes instead of functions.

Functions have their own directory structure, depending on the language, as referenced here: `https://cloud.google.com/functions/docs/writing`. Taking Python as an example, at a minimum, our project directory should contain two files:

- `main.py`: This file, which is at the root of the project directory, will contain the entry point function
- `requirements.txt`: Dependencies for the Python function that use the `pip` tool/standard

If your function doesn't have internet access, you will need to package the dependencies along with your function locally.

More details on dependency management can be found here: `https://cloud.google.com/functions/docs/writing/specifying-dependencies-python`.

Creating your first function

We will now create a Python function. We will use the newer 2nd gen functions as they are going to be the new standard based on the CloudRun platform (which we will cover in an upcoming section).

Since we are using Cloud Shell, Python 3.9 is available to us. If you are experimenting on your own machine, follow the guide at https://cloud.google.com/python/docs/setup:

```
safeercm@cloudshell:~ (packt-serverless101)$ python --version
Python 3.9.2
```

Create a virtual environment for your first function and switch to it:

```
safeercm@cloudshell:~ (packt-serverless101)$ python -m venv gfunction
safeercm@cloudshell:~ (packt-serverless101)$ source gfunction/bin/
activate
(gfunction) safeercm@cloudshell:~ (packt-serverless101)$cd gfunction/
safeercm@cloudshell:~/gfunction (packt-serverless101)$
```

Now, let's create a simple HTTP function that just returns a JSON document:

```
import functions_framework
Import json

@functions_framework.http
def send_message(request):
    data = { "message": "Your first cloud function" }
    return(json.dumps(data))
```

As you can see, send_message is the handler function, which takes a request argument (which we are not processing in this iteration). In HTTP Cloud Functions, the request argument is of the **Flask Request** type. Flask is a lightweight web framework for Python (refer to https://flask.palletsprojects.com/en/2.1.x/api/#flask.Request for more details).

The function entry point is decorated with the HTTP decorator from the Function Framework for Python. You can learn more about this framework at https://github.com/GoogleCloudPlatform/functions-framework-python. Given that we need this framework to operate, this is a functional dependency and should be listed in the requirements.txt file, as follows:

```
functions-framework==3.*
```

Now, let's deploy the function. However, before that, we need to ensure the necessary services are enabled. To do so, run the following commands:

```
safeercm@cloudshell:~/gfunction (packt-serverless101)$ gcloud services
enable cloudfunctions.googleapis.com
Operation "operations/acf.p2-938208393957-a9cf0ae0-b68e-4413-
```

```
b1ed-cb5297400417" finished successfully.
safeercm@cloudshell:~/gfunction (packt-serverless101)$ gcloud services
enable cloudbuild.googleapis.com
Operation "operations/acf.p2-938208393957-dc261d24-1091-4be2-b1e9-
5b1253d1f2f2" finished successfully.
safeercm@cloudshell:~/gfunction (packt-serverless101)$ gcloud services
enable artifactregistry.googleapis.com
Operation "operations/acat.p2-938208393957-86a2a1ad-ec77-4002-aa7a-
7d0fdcf8e156" finished successfully.
```

Also, set your region preferences for functions:

```
safeercm@cloudshell:~/gfunction (packt-serverless101)$ gcloud config
set functions/region asia-south1
Updated property [functions/region]
```

We are going to use the following command to create the function:

```
gcloud functions deploy first-http-function --gen2 --region=asia-
south1  --runtime=python39 --source=. --entry-point=send_message
--trigger-http --allow-unauthenticated
```

Let's look at the arguments (`--gen2` and `--region` are self-explanatory):

- `--runtime`: Each runtime has a unique name (see `https://cloud.google.com/functions/docs/concepts/exec#runtimes` for details).

- `--source`: The function code can be deployed from three locations:

  - Your local development directory

  - A GCS bucket

  - A GitHub repository

- `--entry-point`: As explained earlier, in the context of Python, it is the handler function that will be executed.

- `--trigger-http`: One or more arguments that specify the type of trigger and the trigger type-specific arguments. In this case, the trigger is HTTP (`--trigger-http`) and has an extra argument.

- `--allow-unauthenticated`: This specifies that the resulting HTTP endpoint won't use authentication.

You can also add options to limit the CPU and RAM used by the function; refer to the following document to learn about more useful CLI options – `https://cloud.google.com/sdk/docs/cheatsheet`

For more details on general deployment methods, naming conventions, and more, refer to `https://cloud.google.com/functions/docs/deploy`.

If the deployment fails for some reason, you will be provided with a URL to the log services, along with a query to retrieve the function deployment logs. It will start with something like this: `https://console.cloud.google.com/logs/query`.

Now, let's go ahead and deploy it:

```
safeercm@cloudshell:~/gfunction (packt-serverless101)$ gcloud
functions deploy first-http-function --gen2 --region=asia-
south1  --runtime=python39 --source=. --entry-point=send_message
--trigger-http --allow-unauthenticated
Preparing function...done.
X  Updating function (may take a while)...
  OK [Build] Logs are available at [https://console.cloud.google.com/
cloud-build/builds;region=asia-south1/af7bb952-d363-415f-bb2d-856ed3bd
6818?project=938208393957]
     [Service]
  .  [ArtifactRegistry]
  .  [Healthcheck]
  .  [Triggercheck]
<<OUTPUT TURNCATED>>
  revision: first-http-function-00001-tof
  service: projects/packt-serverless101/locations/asia-south1/
services/first-http-function
  serviceAccountEmail: 938208393957-compute@developer.gserviceaccount.
com
  timeoutSeconds: 60
  uri: https://first-http-function-uzgdrymmva-el.a.run.app
state: ACTIVE
```

A successful deployment will throw a lot of output at you. What is relevant to note down is the HTTP URL that will trigger the function. It can be seen toward the bottom of the output, as a line starting with `uri:`. Other relevant details to note are that the function is archived into a ZIP file and uploaded to a GCS bucket, which is created automatically for us, and the GCP Cloud Build service builds the container with this ZIP file and pushes it to the GCP Artifact Registry. You can read the rest of the output as well; most of the items are self-explanatory.

The URL for the deployed app is `https://first-http-function-uzgdrymmva-el.a.run.app`. Let's hit it and see what the output is:

```
safeercm@cloudshell:~/gfunction (packt-serverless101)$ curl -s
https://first-http-function-uzgdrymmva-el.a.run.app
{"message": "Your first cloud function"}
```

As you can see, we have received the expected output. This is an oversimplified example to illustrate the workflow. To learn more about HTTP functions, refer to the following links:

- `https://cloud.google.com/functions/docs/writing/write-http-functions`

- `https://cloud.google.com/functions/docs/calling/http`

Any function can be invoked directly using the CLI. If your entry point takes an argument, that can be passed as an argument (`--data`), as we will see in the next sections. Let's try invoking this function from the CLI:

```
safeercm@cloudshell:~/gfunction (packt-serverless101)$ gcloud
functions call --gen2 first-http-function
  '{"message": "Your first cloud function"}'
```

Again, the result is as expected. You can fetch the logs of the function using the following command:

```
safeercm@cloudshell:~ (packt-serverless101)$ gcloud functions logs
read –gen2 first-http-function
```

You can combine direct invocation and log viewing for basic troubleshooting. You can list all existing functions using the following command:

```
safeercm@cloudshell:~ (packt-serverless101)$ gcloud functions list
NAME: first-http-function
STATE: ACTIVE
TRIGGER: HTTP Trigger
REGION: asia-south1
ENVIRONMENT: 2nd gen
```

Now that we have deployed our first function and understood its basic building blocks, let's move on to the next important serverless service: GCP Pub/Sub.

GCP Pub/Sub

GCP Pub/Sub is a messaging middleware that allows you to move a large number of events across systems asynchronously. It is called middleware since it sits between data-gathering systems and data-generating systems. It uses a publisher-subscriber pattern to provide asynchronous and scalable message delivery. Publishers are systems that produce messages/events, while consumers are the systems that process the data.

Some of its key features are as follows:

- Global nature – seamless work can be done across geographical regions with global topics
- Large-scale data ingestion capabilities

- Zero provisioning and autoscaling

- End-to-end encryption

- Access control with IAM policies

- Durable message storage

- Retention policies to control storage

- Message delivery guarantees with at-least-once delivery

- HIPAA-compliant

- Attribute-based message filtering

- Seek and replay backlog messages for better failure handling

- Dead letter queues/topics

The pricing model for Pub/Sub is quite simple, with the first 10 GB per month of data flow being free. After that, you are charged a flat $40 fee per TB.

Some of the common use cases for Pub/Sub include user interaction ingestion from client applications, an event bus for enterprise-wide events, reliable failure handling between data processing nodes, change capture data systems, event streaming, and more.

The different types of Pub/Sub flavors

There are two types of Pub/Sub services:

- **Pub/Sub**: This service provides a large number of integrations, high reliability with multi-zone synchronous replication, and autoscaling for any capacity. It is the default choice for most business use cases.

- **Pub/Sub Lite**: This service provides a subset of integrations compared to the regular Pub/Sub service. Scaling is not elastic and customers need to pre-provision capacity in advance. It has low reliability and no synchronous replication

For a detailed comparison before you pick either, visit `https://cloud.google.com/pubsub/docs/choosing-pubsub-or-lite`.

Core concepts

The following diagram provides a quick view of the important concepts in Pub/Sub, along with its basic architecture:

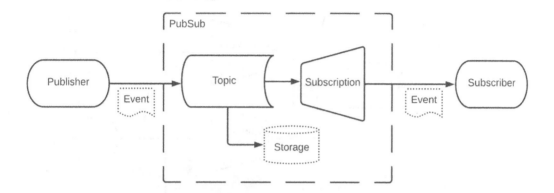

Figure 5.1 – GCP Pub/Sub core concepts

Let's look at these concepts in more detail:

- **Publisher and Subscriber**: As discussed earlier, Pub/Sub stands between message producers and message processors. In the Pub/Sub language, they are called subscribers and publishers, respectively. A publisher or subscriber is a piece of code that runs in the form of different workloads, such as functions, microservices, scheduled jobs, and so on.

- **Topic**: A topic is a named resource to which publishers push their messages. A topic is like a virtual unit of storage for one category of messages and will have capabilities to store and retrieve them.

- **Event/message**: This is the data and optional metadata that publishers send to the topic. It is one unit of information in the Pub/Sub world.

- **Message attributes**: These are one or more key-value pairs that a producer associates with a message. Attributes are usually used to further enrich the data that the message carries.

- **Subscription**: This is another named resource that acts as a proxy to a stream of messages for a single topic. These message streams are intended for one or more subscribers who want to receive messages from the topic.

- **Acknowledgments**: These are short messages/signals that are sent back by a subscriber after it has consumed a message successfully. The acknowledged message will no longer be available with the subscription.

While the preceding diagram depicted the Pub/Sub relationship as a one-to-one relationship, that is not the case most of the time. Pub/Sub relationships can be one-to-many, many-to-one, and many-to-many. The following diagram illustrates this:

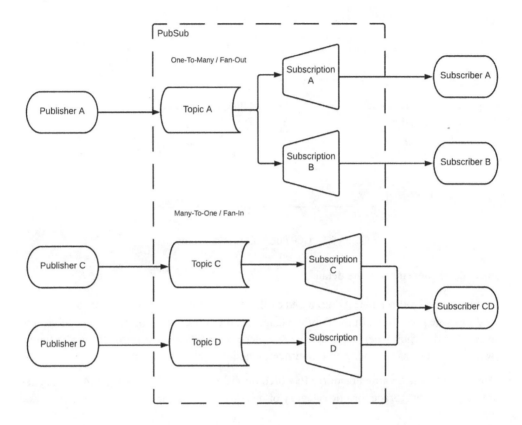

Figure 5.2 – Fan-out and fan-in models

The many-to-many relationship is a combination of the fan-in and fan-out scenarios shown in the diagram.

Another essential feature you need to know about is how subscriptions deliver messages. There are three options:

- **Push**: In this case, Pub/Sub will push messages to the subscribers. This is good for smaller volume traffic but requires that the subscriber has an HTTP endpoint to which Pub/Sub can submit the events.

- **Pull**: In this case, the subscriber initiates pulling the messages from the Pub/Sub platform. This is advisable for high-scale workloads.

- **BigQuery**: This helps in loading the events from a subscription directly to a big query table – which otherwise would have taken the subscriber an extra effort to push the consumed event to BigTable.

To understand more about the subscriber semantics, visit `https://cloud.google.com/pubsub/docs/subscriber`. More detailed coverage of the Pub/Sub architecture can be found at `https://cloud.google.com/pubsub/architecture`.

Now that we have a basic understanding of Pub/Sub, let's do some hands-on experiments. We will start by creating a topic and a subscription:

```
safeercm@cloudshell:~ (packt-serverless101)$ gcloud pubsub topics
create packt-serverless101-topic101
Created topic [projects/packt-serverless101/topics/packt-
serverless101-topic101].

safeercm@cloudshell:~ (packt-serverless101)$ gcloud pubsub
subscriptions create packt-serverless101-subscription101 --topic
packt-serverless101-topic101
Created subscription [projects/packt-serverless101/subscriptions/
packt-serverless101-subscription101].
```

Now, let's produce a message on this topic and consume it through the subscription using the gcloud CLI:

```
safeercm@cloudshell:~ (packt-serverless101)$ for msg in oone two
three;do gcloud pubsub topics publish  packt-serverless101-topic101
--message "message_$msg";done
messageIds:
- '7010178914392397'
messageIds:
- '7010087718131271'
messageIds:
- '7010060739712631'
```

The number that is returned is the message ID of each message. Now, let's consume them. As described earlier, there are three ways to subscribe and consume a message; the only options for us are to push and pull since we are not looking for big query usage. Out of the remaining two, push requires that our consumer stand up an HTTPS endpoint to which Pub/Sub can publish the messages. Since the only option left for us is to pull the message, let's do that.

When consuming in pull mode, each message delivery needs to be acknowledged by Pub/Sub. The gcloud CLI will do that for us if we use the `--auto-ack` flag. Also, by default, `pull` returns only one message, so we will use `--limit 3` to get a maximum of up to three messages from the subscription in one shot:

```
safeercm@cloudshell:~ (packt-serverless101)$ gcloud pubsub
subscriptions pull packt-serverless101-subscription101 --auto-ack
--limit 3
DATA: message_three
MESSAGE_ID: 7010060739712631
ORDERING_KEY:
ATTRIBUTES:
DELIVERY_ATTEMPT:
ACK_STATUS: SUCCESS

safeercm@cloudshell:~ (packt-serverless101)$ gcloud pubsub
subscriptions pull packt-serverless101-subscription101 --auto-ack
--limit 3
DATA: message_oone
MESSAGE_ID: 7010178914392397
ORDERING_KEY:
ATTRIBUTES:
DELIVERY_ATTEMPT:
ACK_STATUS: SUCCESS

DATA: message_two
MESSAGE_ID: 7010087718131271
ORDERING_KEY:
ATTRIBUTES:
DELIVERY_ATTEMPT:
ACK_STATUS: SUCCESS

safeercm@cloudshell:~ (packt-serverless101)$ gcloud pubsub
subscriptions pull packt-serverless101-subscription101 --auto-ack
--limit 3
Listed 0 items.
```

As you can see, `pull` doesn't guarantee that all messages are delivered in one shot, even if you specify the maximum number of messages. Hence, we had to try twice to get all three of our messages.

Let's retry this with a single message:

```
safeercm@cloudshell:~ (packt-serverless101)$ gcloud pubsub topics
publish  packt-serverless101-topic101 --message "message_four"
messageIds:
- '7011203560499945'
```

```
safeercm@cloudshell:~ (packt-serverless101)$ gcloud pubsub
subscriptions pull packt-serverless101-subscription101
DATA: message_three
MESSAGE_ID: 7011250704232523
ORDERING_KEY:
ATTRIBUTES:
DELIVERY_ATTEMPT:
ACK_ID: U0RQBhYsXUZIUTcZCGhRDk9eIz81IChFGwkIFAV8fXFYUXVbVRoHUQ0Zcnxpc2
xfGgRQQlJ2VVkRDXptXG30-fm2RF9BcmpbEgMHRVR-W1kYD21eX12C8bmpr63oSXBmK7qr
sfxIf9iyosduZiA9XxJLLD5-NjxFQV5AEkw9CURJUytDCypYEU4EISE-MD5F
```

As you can see, there is a long string for ACK_ID that Pub/Sub returns for each message that's pulled.

We can do the same programmatically. Since we are dealing with serverless, we should use an example with Cloud Functions as the target of the topic. However, we will do that after we learn about a couple more serverless services; then, we will cover all of them in one example.

To learn more about Pub/Sub, visit https://cloud.google.com/pubsub. Next, we will look at **GCS**.

GCS

GCS is the object store offering from GCP. The units of storage are objects, which GCP defines as *"immutable pieces of data consisting of a file of any format."* Objects are stored in a container, known as a bucket, and are associated with a project. You can associate additional metadata with these objects. Like every other object store, the primary use case for buckets is to upload and download objects.

Some of the key features of GCS are as follows:

- Highly durable, low-latency storage, with an uptime of up to 99.999%.

- Secured in transit and storage.

- Unlimited storage and no object size limit.

- There is no limit to how many buckets you can create.

- There is a limit on how frequently you can create and delete a bucket. This was introduced to avoid abuse.

- Buckets are globally scoped but their contents are stored in a geographic location.

- Bucket names have to be globally unique – across GCP and not just within your account/organization.

- There are several naming convention constraints for GCP buckets. You can find the full list here: https://cloud.google.com/storage/docs/buckets.

- Object versioning is provided, which allows users to see the older versions of an object, as well as recover deleted objects. Keeping older versions comes with the cost of additional storage, though.

- A key feature of GCS objects is storage classes. Technically, these are metadata that is associated with every object. A storage class defines the storage guarantees and pricing of the object. You can associate a default storage class with your bucket, which will be the default for all objects that are uploaded to it. However, it's possible to change the storage class per object. To learn more about storage classes, refer to `https://cloud.google.com/storage/docs/storage-classes`.

You can use life cycle policies to move objects from one class to another, depending on the need and availability guarantee that an object will need during different stages of processing and storage.

GCS is a simple-enough service with high reliability and a variety of use cases. Due to its global nature, it can be used to host websites as well, not just store your images or other objects. We will go ahead and create a GCS bucket and manage an object with it using the gcloud CLI.

> **Note**
>
> There is a legacy utility called **gsutil** that can be used to manage Google buckets. You should use gcloud whenever possible since `gsutil` has some limitations. Visit `https://cloud.google.com/storage/docs/gsutil` to learn more.

Let's go ahead and create a bucket and upload an object to it. Note that all GCS bucket arguments need to be prefixed with `gs://`:

```
safeercm@cloudshell:~ (packt-serverless101)$ gcloud storage buckets
create gs://packt-serverless101-gcs101
Creating gs://packt-serverless101-gcs101/...


safeercm@cloudshell:~ (packt-serverless101)$ gcloud storage
buckets list gs://packt-serverless101-gcs101|grep -E
"id|selfLink|storageClass|location"
id: packt-serverless101-gcs101
location: US
locationType: multi-region
selfLink: https://www.googleapis.com/storage/v1/b/packt-serverless101-
gcs101
storageClass: STANDARD

safeercm@cloudshell:~ (packt-serverless101)$ gcloud storage cp /tmp/
cat.jpg gs://packt-serverless101-gcs101
Copying file:///tmp/cat.jpg to gs://packt-serverless101-gcs101/cat.jpg
  Completed files 1/1 | 191.3kiB/191.3kiB


safeercm@cloudshell:~ (packt-serverless101)$ gcloud storage
```

```
objects describe gs://packt-serverless101-gcs101/cat.jpg|grep -E
"contentType|size|kind|id"
contentType: image/jpeg
id: packt-serverless101-gcs101/cat.jpg/1676746800273688
kind: storage#object
size: '195861'
```

As you can see, we were able to create a bucket, upload a cat image to it, and view the bucket and image metadata.

A very common use case for buckets is to host a public website. Let's create a small HTML file and use the previously created bucket to host the website:

```
safeercm@cloudshell:~ (packt-serverless101)$ echo
'<html><body><h1>Public website from GCS Bucket</h1></body></html>' >
/tmp/index.html

safeercm@cloudshell:~ (packt-serverless101)$ gcloud storage cp /tmp/
index.html gs://packt-serverless101-gcs101
Copying file:///tmp/index.html to gs://packt-serverless101-gcs101/
index.html
  Completed files 1/1 | 66.0B/66.0B
```

Now that we have the HTML file in the root location of the bucket, let's enable public access to the bucket. Be careful with this command and ensure you don't have any private data hosted in the bucket as we are about to make it public. You should also set your index file/main page, as shown in the following command:

```
safeercm@cloudshell:~ (packt-serverless101)$ gcloud storage
buckets add-iam-policy-binding  gs://packt-serverless101-gcs101
--member=allUsers --role=roles/storage.objectViewer
bindings:
- members:
  - projectEditor:packt-serverless101
  - projectOwner:packt-serverless101
  role: roles/storage.legacyBucketOwner
- members:
  - projectViewer:packt-serverless101
  role: roles/storage.legacyBucketReader
- members:
  - allUsers
  role: roles/storage.objectViewer
etag: CAI=
kind: storage#policy
resourceId: projects/_/buckets/packt-serverless101-gcs101
version: 1
```

```
safeercm@cloudshell:~ (packt-serverless101)$ gcloud storage buckets
update gs://packt-serverless101-gcs101 --web-main-page-suffix=index.
html
Updating gs://packt-serverless101-gcs101/...
  Completed 1
```

While we have completed the steps to serve the content from our bucket, actually serving them to a user requires you to set up either a custom domain or GCP load balancer for the bucket. Both are outside the scope of this topic, so let's just view the file by hitting `index.html` directly:

```
safeercm@cloudshell:~ (packt-serverless101)$ curl https://storage.
googleapis.com/packt-serverless101-gcs101/index.html
<html><body><h1>Public website from GCS Bucket</h1></body></html>
```

To learn more about website hosting, visit `https://cloud.google.com/storage/docs/hosting-static-website`. Next, we will look into Cloud Workflows.

Cloud Workflows

Cloud Workflows is an orchestration service that GCP introduced in 2020 and made generally available in 2021. Like other serverless offerings, this is also a fully managed solution with a pay-as-you-go model. Before Cloud Workflows, the managed workflow offering that GCP provided was called Cloud Composer – based on the open source Apache Airflow.

Cloud Workflows can orchestrate the interaction between several Google Cloud services. A workflow is a series of steps that specifies the tasks to be performed at each step, and how to move to the next step in the workflow based on the conditions or output of the previous job. These workflows are usually expressed either in YAML or JSON format.

Some of the key features of Workflows can be seen here:

- **Reliability**: The workflows are replicated across multiple zones in a region for redundancy. After the execution of each step, the workflow's state is checkpointed so that in the case of outages, they can start right from where they left off.

- **Dependency failure handling**: Default and custom retry policies, timeouts, and custom error handling.

- **Long-running operations**: Workflows support long-running operations with a waiting period of up to 1 year.

- **Variety of triggers**: Workflows can be kicked off by events – using the Eventarc service – by schedules, or programmatically using SDKs.

- **Callbacks**: HTTP callbacks can be created inside the workflow so that external systems or humans can provide input between different steps of the workflow. Combining this with support for long-running operations makes it a powerful tool to automate a lot of workflows that need human intervention or wait for an external system to process something in between the steps.

- **Low latency execution**: Wofkows have no cold start delays, which means they start as soon as you/or an external trigger kicks them off.

- **Developer experience**: Supports fast deployments, integrated logging, and monitoring for better debugging.

- **Security**: Sandboxed execution of workflows, and integration with Google Cloud IAM/Service Accounts for authentication and authorization.

These workflows are priced based on many parameters, with the primary one being the number of steps in the workflow. At the time of writing, the first 5,000 steps – across any number of workflows and their invocations – are free. If any steps make external HTTP calls, the first 2,000 are free.

Common use cases

A large number of business use cases can be represented as workflows. In many cases, such business use cases involve invoking a series of purpose-built microservices that execute a specific function. This microservice orchestration can be carried out by workflows. Service integration is another use case where multiple services can be connected by using workflows to process the output of one application or service and invoke another service as a response or for further processing.

IT process automation is a very common use case. Consider the example of a new employee joining where their laptop, security badge, laptop bag, and swag have to be issued by various departments, their email and network access account must be created by IT, and their payroll information must be created by the finance and accounting team. This is partially a combined IT and business workflow. There are also other pure IT workflows, such as provisioning software or creating SaaS accounts.

Data science and AI are other places where workflows are widely used. In these domains, a lot of data movement is involved between various systems. These are typically referred to as data pipelines and need workflows to be defined for the movement, cleanup, and enrichment of data.

There are many more use cases we can find around our businesses and operations, but this section should have given you a good idea of what you can find. Next, we'll look at the format and syntax of a workflow. While workflows support both YAML and JSON, we will stick to the YAML format since it is more human-readable.

Writing a workflow

When you are writing a workflow in the workflow syntax, this is called the workflow definition; the result will be a file in YAML or XML format. The workflow definition must be deployed, which will create the workflow (but won't execute it without a trigger). If you modify the definition, you can redeploy your workflow. An execution is a single run of the workflow and is chargeable. Multiple executions (parallel or serial) of the same workflow are independent of each other. As discussed earlier, workflows can be triggered in several ways.

Let's look at the key elements of the workflow syntax:

- **Steps**: A step is one unit of execution within a workflow. A workflow should have at least one step. Unless specified otherwise, all steps in a workflow are executed in the order they are listed in the definition. For any action that you need to take, a step supports calling an HTTP endpoint. In addition, this step can assign a variable, control execution flow based on conditions, return a value, or pause a workflow. Steps can also be nested.

- **Conditions**: Use the `switch` keyword to evaluate conditions and specify which step to take next.

- **Iteration**: These are like loops in programming. Here, you use a `for` loop to go over a sequence or a collection of data.

- **Parallel**: As mentioned when we talked about steps, by default, steps are executed sequentially. If you want to parallelize any work, you need to use parallel steps.

- **Subworkflows**: A workflow definition needs a minimum of one workflow, called the main workflow, but you can have sub-workflows that you can invoke from the main workflow. This is like writing a function for a task that keeps repeating.

As a word of caution, while workflows can call Google services and manipulate their outputs or inputs, it has limitations – for any complex business logic, implement them in a function or cloud run and invoke them from the workflow.

The workflow language in itself is a mini DSL and can't be fully covered here. A detailed syntax overview can be found here: `https://cloud.google.com/workflows/docs/reference/syntax`. Next, we will explore an example to see how workflows work.

A sample workflow

First, enable the workflow API:

```
safeercm@cloudshell:~ (packt-serverless101)$ gcloud services enable
workflows.googleapis.com
Operation "operations/acat.p2-938208393957-697c4914-0761-4d2f-a964-
6e0c2e700872" finished successfully
```

While enabling this API helps you create the workflow, you need to have logging enabled for your project via a Service Account so that the logs of the workflow's execution can be sent to the GCP log service:

```
safeercm@cloudshell:~ (packt-serverless101)$ gcloud iam service-
accounts create sa-name
Created service account [sa-name].
safeercm@cloudshell:~ (packt-serverless101)$ gcloud projects add-iam-
policy-binding packt-serverless101 \
    --member "serviceAccount:sa-name@packt-serverless101.iam.
gserviceaccount.com" \
    --role "roles/logging.logWriter"
Updated IAM policy for project [packt-serverless101].
bindings:
<<TRUNCATED OUTPUT>>
version: 1
```

To learn more, visit https://cloud.google.com/workflows/docs/create-workflow-gcloud.

Now, let's create a simple workflow that will call an API, process the data, and output the result.

What we are going to do is a simple case of converting a certain amount of money in one currency into another currency based on the current exchange rate. The amount and the currency to be converted to will be arguments to the workflow, while the base currency will always remain USD. First, we will look at the entire workflow code; then, we will describe each step. Given that this is a simple use case, we will only have a main workflow and no sub-workflows:

```
main:
    params: [inputs]
    steps:
      - getConversionRate:
          call: http.get
          args:
            url: https://api.currencyapi.com/v3/latest
            query:
              apikey: XWM9QxuQ32qLEgdfN9aqIzZzAYE2D6mPZ9RQlEEg
              base_currency: USD
              currencies: ${inputs.currency}
          result: convRateResult
      - logger:
          call: sys.log
          args:
            text: inputs.amount
      - convertAmount:
          assign:
```

```
                - rate: ${convRateResult.body.data[inputs.currency].
     value}
                - base_amount: ${ inputs.amount }
                - converted_amount: ${ double(rate) * int(base_amount) }
       - returnAmount:
            return: ${converted_amount}
```

Now, let's look at each step and understand what they do. These steps can be found under `main | steps`; each has a different name.

getConversionRate

This step makes a call to `currencyapi.com` to get the current conversion rate between the base currency (USD) and the currency to be converted into (which will be an argument to the workflow, along with the dollar amount to be converted). Sign up with `currencyapi.com` to get your own API key and replace the API key value in the workflow definition. As you can see, an HTTP call is made to the API with two query parameters, one of which is the API key; the other is the conversion currency. As a good practice, the API key should be stored in Google Secret Manager, but for now, we will embed it into the code. For best practices on this, refer to `https://cloud.google.com/workflows/docs/use-secret-manager`.

There are a couple of other things to notice in this step:

- The result of the previous step was stored in `convRateResult`.
- `inputs` is the variable where your workflow input is stored as a dictionary. Refer to `https://cloud.google.com/workflows/docs/passing-runtime-arguments` for more details.
- We are making an HTTP call in the step (`https://cloud.google.com/workflows/docs/http-requests`). You can learn more about calls in general at `https://cloud.google.com/workflows/docs/reference/syntax/calls`.
- `https://currencyapi.com/` provides a free API key, but you need to sign up for it.

convertAmount

In this step, we just use basic math to convert the amount using the workflow constructs of variables and expressions. We also use some of the standard library functions to help with converting data types:

- Expressions are for evaluating something such as a calculation, variables, inputs, and so on (ref: `https://cloud.google.com/workflows/docs/reference/syntax/expressions`)
- The standard library offers a ton of helpful functions (ref: `https://cloud.google.com/workflows/docs/reference/stdlib/overview`)

returnAmount

As its name suggests, this step returns the amount that was calculated in the previous section. To learn more about steps and what they can do, refer to https://cloud.google.com/workflows/docs/reference/syntax/steps.

Now, we will create the workflow with the gcloud CLI and deploy it:

```
safeercm@cloudshell:~ (packt-serverless101)$ gcloud workflows deploy
usd_convert --source=usd_convert.yml     --service-account=sa-name@
packt-serverless101.iam.gserviceaccount.com
<<OUTPUT TRUNCATED>>
name: projects/packt-serverless101/locations/us-central1/workflows/
usd_convert
revisionId: 000001-daf
<<OUTPUT TRUNCATED>>
state: ACTIVE
```

Next, we'll call the workflow with our input:

```
safeercm@cloudshell:~ (packt-serverless101)$ gcloud workflows run usd_
convert --data '{"amount":1001,"currency":"EUR"}'
<<OUTPUT TRUNCATED>>
result: '944.495552'
state: SUCCEEDED
<<OUTPUT TRUNCATED>>
workflowRevisionId: 000001-daf

safeercm@cloudshell:~ (packt-serverless101)$ gcloud workflows run usd_
convert --data '{"amount":1001,"currency":"INR"}'
<<OUTPUT TRUNCATED>>
result: '82700.921303'
state: SUCCEEDED
<<OUTPUT TRUNCATED>>
```

This concludes our section on Cloud Workflows. This brings us to the end of the serverless services we will cover in detail in this chapter. The following section will give you a sneak peek into what other serverless services GCP provides.

More serverless services

The following services are useful when combined with the services we've covered in this chapter. I will try to cover them in order of importance, though that order might be different based on what you are trying to solve.

Cloud Run

This service can be considered the big cousin of Cloud Functions. Cloud Run offers more power and flexibility than Cloud Functions for more complex use cases. It is a fully managed serverless container platform. It allows you to bring in custom containers with your own runtimes and custom libraries. This way, you are not limited to the choice of OS and runtimes that Cloud Functions offers. You can also do your own optimizations on the runtimes and OS so that they run the way you want.

A detailed write-up on when to use Cloud Run versus Cloud Functions can be found at `https://cloud.google.com/blog/products/serverless/cloud-run-vs-cloud-functions-for-serverless`.

Eventarc

Eventarc manages the event flow from sources to targets without the customer having to worry about the integrations or underlying infrastructure. The sources can be most of the Google Cloud services (130 at the time of writing) or many third-party services. These sources are called providers in the Eventarc system. Eventarc is responsible for routing these events based on configured rules to the destinations, otherwise known as event receivers/consumers. It supports the common CloudEvent format (`https://cloudevents.io/`). At the time of writing, Eventarc supports Cloud Functions (v2), Cloud Run, Workflows, and **Google Kubernetes Engine (GKE)** as receivers.

To learn more about Eventarc, visit `https://cloud.google.com/eventarc/docs/overview`.

Cloud Scheduler

If you are familiar with `crontab` in Linux, this should be a very familiar service to you. Cloud Scheduler, as its name suggests, works as a cloud-native job scheduler. It is built for resilience and unlike `crontab`, you can configure retries for all the tasks you schedule. You can use it as the central place from which you can organize all your cloud automation tasks.

Refer to `https://cloud.google.com/scheduler` to learn more.

Databases and data stores

The following list specifies several data services that can be used in conjunction with serverless in GCP:

Cloud BigTable	A NoSQL store for key-values
Datastore	A NoSQL store with indexing
CloudSpanner	A global-scale RDBS similar to AWS Aurora
CloudSQL	To manage MySQL, PostgreSQL, and more, similar to AWS RDS
BigQuery	A serverless data warehouse service

Table 5.2 – Important GCP data stores

Now that we have gone through some of the core serverless GCP services, let's look at the project that we will implement in this chapter. But before we do, make sure you delete all the resources we created in this chapter. Usually, you can use the `gcloud <service> delete <options> <resourcename>` command to do this. Use `gcloud help` to find the right syntax for each. Now, let's look at the project.

The project – nameplate scanning and traffic notice

It is very common for cameras to capture speeding vehicles. As a consequence, the owners get traffic notices to pay a fine. We will automate a trimmed-down version of this process using GCP serverless products:

1. The captured number plates will be uploaded to a Google Cloud storage bucket.

2. An Eventarc rule will be triggered due to this upload event and will subsequently invoke a Cloud Run container.

3. The Cloud Run container will process the image and extract the number plate, which will be sent to a Pub/Sub topic.

4. The Pub/Sub message will trigger a cloud function that will pick up the owner's details from a Cloud Spanner database, generate a PDF report, and upload it to another cloud bucket for further processing.

The architecture will look as follows:

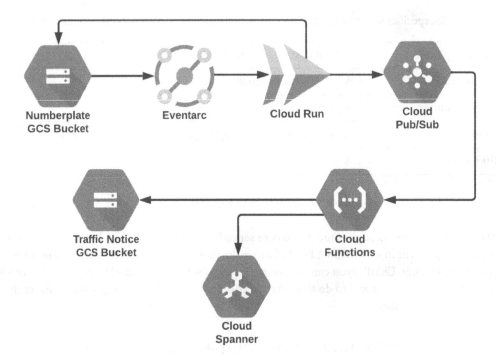

Figure 5.3 – GCP project – nameplate scanning

We have provided the implementation for this in this book's GitHub repository at `https://github.com/PacktPublishing/Architecting-Cloud-Native-Serverless-Solutions/tree/main/chapter-5`. With this, we have come to the end of this chapter.

Summary

We started this chapter by covering the basic constructs of GCP and how to set up your account and start developing with Cloud Shell. We then covered the most important serverless service, Cloud Functions. This was followed by Pub/Sub and Google Cloud Storage. Then, we moved on to workflows, where we learned about the workflow language and how to construct a basic workflow. We also took a quick look at some other serverless services in short. Finally, we concluded with a project that you will implement with the knowledge you've gained from this chapter. You will also need to use a couple of more extra services that we didn't experiment with in this chapter.

Overall, this chapter provided you with a good idea of what GCP serverless has to offer. In the next chapter, we will look at a non-conventional serverless provider – Cloudflare – that is not exactly a cloud vendor.

6
Serverless Cloudflare

In the serverless world, Cloudflare stands apart from the cloud providers we have explored in previous chapters. Cloudflare doesn't offer a cloud platform or even a comprehensive serverless platform suite. Instead, it offers services that are aligned with its core business.

Cloudflare started in 2009 by offering protection from email spam, but it soon expanded its portfolio into an array of network and content-related services. Most people initially knew Cloudflare as a **Content Delivery Network (CDN)** and a **Domain Name Service (DNS)** company. But it is much more than that. Today, Cloudflare boasts an array of internet services that fall into the categories of application and network security, web performance and reliability, collaboration and zero-trust security, and serverless platforms. Each of these categories provides several services under it.

The biggest advantage Cloudflare has is its vast and powerful network infrastructure. At its core, Cloudflare is an HTTP-based web caching network spread across 250+ edge locations. To become a top content delivery network provider, Cloudflare had to build a large number of PoPs and interconnect/peer with a lot of **internet service providers (ISPs)** across the globe. Being a DNS provider added to its capabilities as it gave them an easier way to influence global DNS service partially and most of its customer traffic fully. This level of control helps them implement **Global Service Load Balancers (GSLBs)** based on DNS effectively. Securing this infrastructure and its customers in a performant way allowed Cloudflare to innovate further and come up with an array of security and performance-related services. Serverless offerings came out from one such innovation. In this chapter, we will look at the serverless platforms offered by Cloudflare and how it works well with the rest of its services.

In this chapter, we are going to cover the following topics:

- Cloudflare service portfolio
- Cloudflare Workers – the workhorse at the edge
- Cloudflare Workers KV
- Cloudflare Pages
- Newest Cloudflare serverless offerings

- Workers and KV – learning by example
- Project
- More vendors in edge computing and JAMStack

Cloudflare service portfolio

The following table provides a categorized list of the services provided by Cloudflare:

Applications and Network Security		
API Shield	Bot Management	DDoS Protection
Magic Transit	Magic WAN	Network Interconnect
Rate Limiting	TCP/UDP Applications	SSL/TLS
SSL/TLS for SaaS	Web Application Firewall	
Performance and Reliability		
Argo Smart Routing	CDN	China Network
DNS	Load Balancing	Stream Delivery
Waiting Room		
Zero-Trust Network		
Cloudflare for Teams	Access	Gateway
Browser Isolation		
Serverless Solutions		
Workers	Workers KV	R2
Durable Objects		
Website Development		
Pages	Streams	
End User Network/VPN		

1.1.1.1	1.1.1.1 with WARP	1.1.1.1 for Families
Analytics and Insights		
Analytics	Cloudflare Logs	Web Analytics
Cloudflare Radar		
Privacy		
Data Localization	Compliance	
Other		
Cloudflare for SaaS	Cloudflare Registrar	

As you can see, the services that Cloudflare provides are themed around content creation and delivery, reaching the end users, and improving last-mile connectivity, stability, and security. Despite having so many services, Cloudflare's core strength is its CDN and Edge Network. Almost all services are built on top of that. To recall what was discussed in *Chapter 2*, the following is a diagram of Cloudflare CDN:

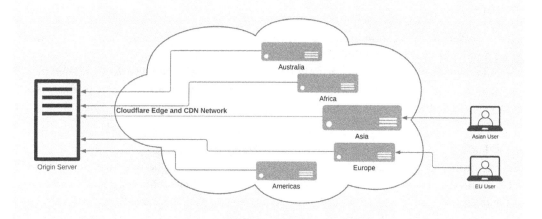

Figure 6.1 – Cloudflare Edge Network

The preceding diagram shows the Cloudflare Edge Network with data centers across the globe and customers connecting to the nearest edge servers from their location. The origin servers of the customer are outside the Cloudflare network – anywhere in the world. Now that you have an idea about the Cloudflare service portfolio and its infrastructure, let us look at the key service that powers the Cloudflare serverless platform – Workers.

Cloudflare Workers – the workhorse at the edge

Cloudflare Workers provides **Function-as-a-Service (FaaS)** at the edge of its network. Workers was introduced in 2017 September as beta and opened up to everyone in 2018 March. Adding programming capabilities to a cache is nothing new. Multiple open source projects support adding programmability to a caching or edge server. The varnish cache (`https://varnish-cache.org/`) provides a configuration language called **varnish configuration language (VCL)**. Similarly, Nginx supports adding LUA programs as modules (`https://www.nginx.com/resources/wiki/modules/lua/`), as do many other projects, including HAProxy, Kong API Gateway, and others. All these services aim to achieve one thing: intercept the HTTP requests and responses at the edge and modify them with custom logic for better performance and/or additional feature support.

Cloudflare and other CDN/edge service providers use different tools and services to keep adding features at the edge. But as Cloudflare pointed out in their blog introducing Workers, it is never going to be enough, as each customer would have their own custom use case and no provider can cover all of those use cases. So, the next best thing that could be done was to liberate the users – by providing them with a platform that could add custom programming logic to the customer endpoints. That is how Cloudflare Workers was born.

Service Workers – the namesake and power behind Cloudflare Workers

The Web Workers API is a common standard implemented by most mainstream browsers. It allows running a script as a background thread while the main execution thread handles interactions with the web services. This allows time-consuming and heavy processing to be performed separately, allowing the main thread to operate without being blocked or slowed down. There are three types of Workers – dedicated, shared, and service. We are interested in the service Workers.

A service worker is a script that the browser runs in the background, and can facilitate a bunch of features that don't need user interaction or web pages. Some of these use cases are push notifications and background sync. Service Workers can act as proxy servers that sit between web applications, browsers, and the network. This feature allows for creating a better offline experience, as well as intercepting requests and responses and taking action on them. Workers provide complete control to developers to improve the user experience. Some of the key features of service Workers are as follows:

- They are JavaScript Workers
- They are programmable network proxies that provide complete control of network requests that a web page makes
- They cache request and response pairs using the Cache API for offline use
- They push notifications
- They perform a background sync

The service worker is a process, and has a life cycle that goes like this:

- **Registration**: Done through the web page's JavaScript
- **Installation**: Download and cache the JavaScript files required
- **Activation**: Activate the worker and start handling events

Being able to intercept traffic and modify request responses makes service Workers very powerful, as well as dangerous. For this reason, service Workers are allowed only on HTTPS-secured web pages.

Traditional web request interception at the edge was based on installing hooks at different stages of the HTTP state machine – when the request is received, the response is sent out, when a response is cached, and so on. With Workers, it installs an event handler against an endpoint. This handler then intercepts the request and responds to it asynchronously. This allows the Workers to do asynchronous I/O before the response is sent out. It can also make requests to other web endpoints (called sub-requests) if needed. Subrequests can be executed in parallel or sequence and then the results can be combined.

Cloudflare Workers is based on JavaScript and service worker technology. It runs on top of the Javascript V8 engine. The V8 engine has a good sandboxing mechanism that makes it a very stable choice for running untrusted code in isolated environments. This is much more lightweight compared to running heavyweight containers. Besides, JavaScript is ubiquitous – it is used on clients and servers and is a highly adopted and widely accepted technology, making it an ideal choice of language for faster customer adoption.

Cloudflare Workers – functionality and features

While Cloudflare Workers uses service Workers as its underlying technology, unlike the service Workers running on the browser, Cloudflare Workers runs on their edge servers. Any request that is received for a domain hosted in Cloudflare can be intercepted and modified by the Workers. When receiving a request, they can either directly respond to the request on their own or forward it to an origin server (backend). In either case, Workers can run their own programming logic and modify the request (before it is sent to the origin servers) or the response (before it is sent to the client browser). Similar to the service Workers, Cloudflare Workers can also work asynchronously. This is useful when some post-processing needs to be done after the response has already been sent to the user, such as via request or response logging or updating analytics services.

The following diagram explains how Cloudflare Workers works:

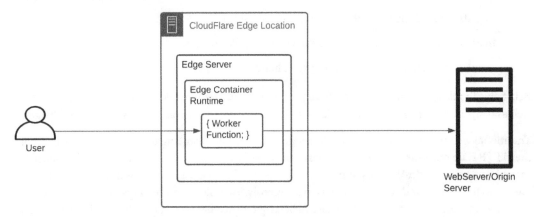

Figure 6.2 – The Cloudflare Workers request-response flow

The following diagram shows how Workers can intercept a request and either respond directly or forward it to an origin server:

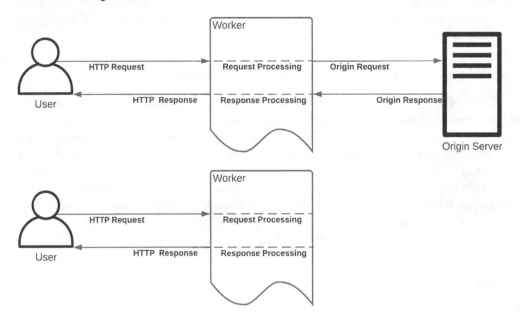

Figure 6.3 – Request intercept – Forward versus respond directly

Some of the typical use cases for Cloudflare Workers are as follows:

- An ecosystem to develop and host serverless applications using REST APIs
- Route traffic to different backend/origin servers based on URL paths, domains, and a whole lot of other parameters
- Provides access control on the edge cache
- Implement canary traffic routing
- Provides custom logic to fail over and load balance origins
- Provides traffic filtering at the edge with custom rules
- Performs A/B testing between different versions of applications running on different backend clusters
- Performs analytics at the edge to gauge customer traffic
- Improvises cache hits by rewriting the incoming requests
- Applies user-facing production changes at the edge so that core backend code doesn't have to be modified
- Uncompresses the web content received from the origin at the edge to save bandwidth costs for the customer

Speed is one of the key features that Cloudflare Workers boasts of. It claims the Workers typically take less than a millisecond. In many cases, this is found to be true, and even when it takes more than that, the advantage of servicing/modifying the request from the edge location improves the overall performance and delivery. The performance of Workers comes from the V8 Javascript engine, as we covered earlier in this chapter. V8 was created by Google for the Chrome browser and is one of the faster implementations of the JavaScript engine. Its speed also applies to the worker's deployments – a worker, when pushed to the edge, is deployed to the global edge locations of Cloudflare in less than 30 seconds. This is an amazing feat.

We talked about the process isolation and sandboxing that the V8 engine offers to Workers. This is made possible with V8 Isolates. Isolates are lightweight contexts that provide the JavaScript code with a safe environment for execution. The worker function runs within this Isolates sandbox. A single JavaScript runtime can run a large number of Isolates and switch processing between them. While the runtime keeps switching between Isolates, the memory of individual Isolates is fully isolated and hence the user code is protected and never interacts with the JavaScript runtime or the neighboring Isolates. A common problem with container-based FaaS platforms is the cold start issue. This is the initial delay in spinning up a new container and its runtime to run the FaaS code. Isolates eliminates this as the cold start happens only once for a runtime environment; then, thousands of Isolates can be run on top of the runtime with virtually no cold starts. While a large number of Isolates can be run in an edge container runtime, there are rare occasions where an Isolate can be stopped and evicted from a runtime.

This could include resource limitations on the machine that the edge container is running on, customer resource limits, or malicious worker code. The following diagram depicts how Isolates are packed compared to containers:

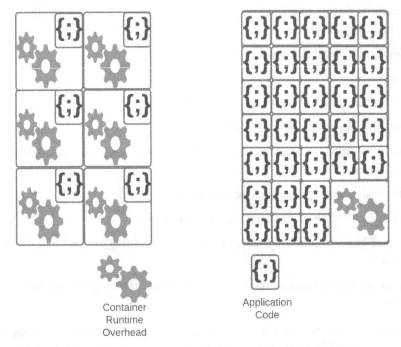

Figure 6.4 – Traditional container-based FaaS versus Isolates

Now that we know how Isolates are organized and how they improve the resource utilization and performance of Workers, let us look at the language support in Workers.

Other languages supported

While Workers is built around the JavaScript V8 engine, it can support other language runtimes. This means you can write your worker code in any of the supported languages, not just JavaScript. Workers started by supporting writing workers in JavaScript, which was the obvious choice due to the platform on which it is built. This was followed by support for languages that compile to **WebAssembly (Wasm)** such as Rust/C/C++. Wasm is a portable binary code format, along with its text format, and can run in modern web browsers just like JavaScript. The Wasm programs are usually small and blazing fast. Usually, you don't write programs in Wasm, but rather write them in higher-level languages such as Rust/C/C++ and then compile them into Wasm format. Some languages can also be compiled into JavaScript. This is a long list and Workers can support most of it. Please refer to the following web page to learn more about the language support of Workers: https://developers.cloudflare. com/workers/platform/languages.

Before we start playing around with Workers, let us look at the next serverless service from Cloudflare – Workers KV.

Cloudflare Workers KV

Workers KV is a key-value store offering from Cloudflare. This store is globally distributed and provides low latency access to the key values. Worker KV is tuned for high-volume reads while supporting relatively low-volume writes. The more data is read, the better performance it provides. The KV architecture ensures that frequently accessed key values are pushed to and accessible from all edge locations while infrequently read key values are stored centrally in limited data centers.

KV is an eventually consistent data store. This means that a modification made at a specific edge location will reflect the change immediately locally but will take time to propagate to all edge data centers. As per the documentation, the ideal time for propagation is 60 seconds, but this could be longer for locations where the same key was accessed recently (and an older version of the value was returned). This is due to the caching nature of the key values at the edge. The old value might be cached at that edge location, and the cache entry might take a bit more time to invalidate.

Due to this eventual consistency model, KV is not ideal for use cases where strong data consistency is sought. It cannot support atomic operations and read-after-write (where a data item is read immediately after it is written) are not a good use case for KV. The KV data is encrypted and secured at rest, as well as in motion. At rest, Cloudflare uses 256-bit AES encryption (which only workers can decrypt) to protect KV; while at motion, **transport layer security** (**TLS**) protects the data.

Next, let us examine CloudFlare Pages, a new way to host static and dynamic websites.

Cloudflare Pages

Cloudflare Pages is a service that can host JAMStack-based frontend applications. It is not just a hosting platform, but rather a complete **Continuous Integration/Continuous Delivery** (**CI/CD**) solution that allows developers to experiment with their code and push it to production with ease. Before we learn more about Pages, let us understand what JAMStack is.

JAMStack

JAMStack is a software development philosophy and architecture that has revolutionized building frontend applications. JAM in JAMStack stands for JavaScript, APIs, and Markup. It was created and popularized by the company Netlify, which uses JAMStack for its web hosting platform.

The distinctive advantage of JAMStack is that it's easier, safer, and more efficient to deliver static web pages that are pre-created than to use dynamically generated web pages based on server-side rendering. The idea of generated websites was popularized by a bunch of frameworks and tools that fall into the category of **Static Site Generators (SSGs)**. Some of the most popular frameworks are Jekyll, GatsbyJS, and Hugo. SSGs can generate a fully static website (based on HTML, CSS, JavaScript, and image/media files) from the raw data provided and a bunch of templates. SSGs can automate the task of generating each HTML page in a website and since these pages are pre-built instead of rendered when accessed, they can load very quickly and outperform traditional dynamic websites.

Traditionally, this kind of content generation was done by **content management systems** (**CMSs**). They would do this dynamically based on the templates, content, and layout defined by the users. SSGs make this process much more simplified. A list of popular SSGs can be found at `https://JAMStack.org/generators/`.

While SSGs solved the problem of page generation and improved performance, the static nature (as intended) limited SSG-generated sites from bringing any dynamic features. Many developers felt that these sites did not have to remain static and from that thought, JAMStack was born. Let us examine the relevance of each technology in JAMStack.

JavaScript

Javascript handles all dynamic functionalities of JAMStack websites. You could use plain vanilla JavaScript or any other framework of your choice. JAMStack does not impose any limitations on your choice of framework.

APIs

Server-side features required by the website are packed into reusable APIs and are hosted anywhere on the internet. These could be custom APIs created by the developers themselves, or hosted services that provide various backend services.

Markup

The static web pages have to be generated from source files and templates. The source files are usually stored as Markdown files. The templates make use of the data defined in the structured language that Markdown provides. Similar to JavaScript, there is no restriction as to which Markdown engine you should use. So long as your SSG supports it, you are good to go.

The SSG structure can be loosely represented as follows:

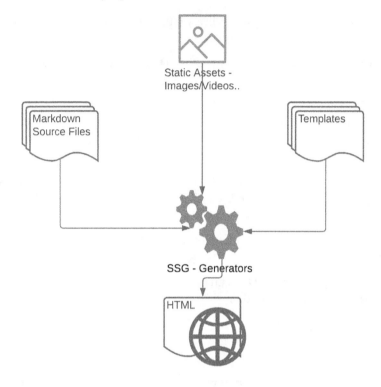

Figure 6.5 – Static Site Generators

The primary advantage of using JAMStack is its performance due to serving static content rather than dynamically generated pages. Serving static content is always faster since dynamic websites have to generate the content to be served and this adds latency to the response. But that is not all – having the content as static files means the entire site can be distributed using CDN, which takes it to locations across the globe. With dynamic sites, the application servers can only be hosted in a limited number of locations, and they can be much farther from the user compared to the CDN edge locations. This reduces the time it takes the user to reach the website.

So far, we have covered how to generate static sites and how it helps improve performance. Another tenant of the JAMStack philosophy is the adaptation of complete CI/CD principles. A JAMStack site is created out of the data files and templates using a static content generator. These source files – templates, data, and so on – are committed to a Git repository. Whenever a change is made to a file and pushed to the repo, a build service runs the site generator, which produces the static pages. These are then deployed to a CDN. This helps in automating the entire developer workflow as there are no servers or infrastructure to manage and the only place where developers interact with this infrastructure is their source code repository.

Some of the advantages of using JAMStack are as follows:

- **High scalability**: Since the hosting and delivery is through CDNs, the scalability is only limited by the CDN edge networks, which are usually quite wide – if you choose wisely, that is.

- **Faster page load**: JAMStack offloads most of the functionalities that servers traditionally used to execute to the browser. By design, it also has limited backend and database calls to make and very little dynamic content to generate.

- **Full-fledged CI/CD adoption**: JAMStack developers practice Git-based CI/CD workflows that improve the developer experience, as well as provide faster software delivery.

- **Better security**: Dynamically rendered web applications expose more attack surfaces that can be exploited. Having a statically generated site reduces that. The lack of database and backend services decreases the threats further. Even when the backend services are being used, they are usually highly secured and (already) acting as publicly exposed endpoints.

- **Better user experience**: This is the direct result of fast page loads, CDN performance, and improved security.

- **Search engine optimization (SEO)**: With full control over the content generation, it is easier to tune for search engine optimization.

- **Portable framework and provider choices**: JAMStack does not enforce specific frameworks or hosting providers. It is quite easy to switch between frameworks and hosting/CDN providers based on convenience and cost factors.

- **Maintainable**: So long as the code that is pushed to Git is stable, you don't have to worry about the upkeep of the website 24/7.

- **Easier to test and debug**: With its full-fledged CI/CD adoption, it is easier to tune your deployment strategy based on the Git workflows and branching strategy the developers follow. This allows you to push code to additional environments for **user acceptance testing (UAT)**, A/B testing, feature flags, and more.

The performance difference between traditional server-side rendering and JAMStack is depicted in the following diagram:

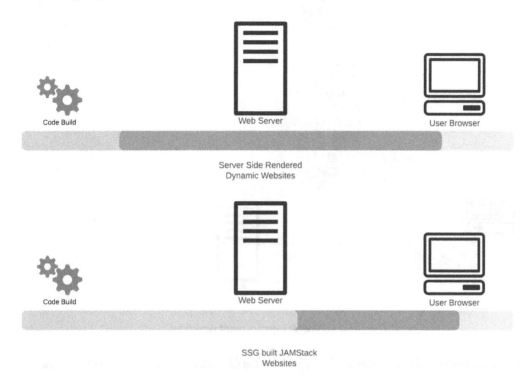

Figure 6.6 – Dynamic websites versus JAMStack

A typical JAMStack workflow involves committing source files to Git, which will trigger a continuous integration system that will run an SSG to create the static website. The continuous delivery system will pick up the output of SSG (the website) and deploy it to the CDNs. When the user accesses the website, dynamicity can be brought in by the JavaScript on the pages calling various APIs – internally developed as well as vendor provided. The following diagram shows what this workflow looks like:

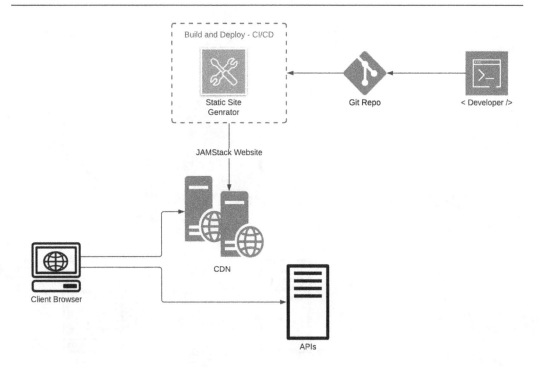

Figure 6.7 – JAMStack deployment workflow

Now that we understand how JAMStack works, let us look at Cloudflare Pages as a platform for JAMStack.

Cloudflare Pages and JAMStack

Cloudflare Pages allows you to build and deploy any JAMStack website. Pages lists several frameworks for website generation (SSG) and can also be used with unlisted frameworks. For implementing CI/CD of code, Pages supports integration with GitHub and GitLab. At the time of writing, these are the only Git platforms supported. It provides a build environment and builds configurations that can be used to customize the build commands and output directory (for static site generators using various tools and frameworks). Pages allows users to deploy to different endpoints and based on the Git branch it is pushed to. This helps developers deploy to test environments and promote production after testing the endpoint corresponding to the test branch.

In addition to the Git integration and deployment, there are a few more customization features that Pages supports. Some of those are as follows:

- **Deployment hooks**: Hooks allow page deployments to be triggered, but external events use webhooks.

- **Preview deployments**: Changes pushed to the non-production branch generate a new instance of the production website with the custom URL. This allows **user acceptance testing** (**UAT**) and other tests.

- **Caching**: Pages provides built-in caching, and the default configuration is optimized for caching as much as possible.

- **Rollback deployments**: Page deployments can be rolled back quickly.

- **Redirects**: Supports configuring various routes to redirect to alternate URLs.

- **Custom Headers**: Configure Pages so that you can add custom headers per page.

- **REST API**: Cloudflare Pages has a REST API endpoint that can be used to configure Pages better.

- **Functions**: Pages has very good integration with Cloudflare Workers and can be used to add dynamic features to the site.

This concludes the production-ready serverless services from Cloudflare, but there are a couple of more services in the works. These products are in private beta and only limited knowledge is available about them. Let us take a brief look.

Newest Cloudflare serverless offerings

Part of Cloudflare's vision is to become an all-rounder in providing a serverless platform at the edge. While Workers, Pages, and KV are the first step toward this vision, they have much more in store. We will examine two products that have been announced and are in private beta already.

Cloudflare R2 storage

R2 is the object storage offering from Cloudflare. Like the other object stores we discussed in previous chapters, R2 can also host large amounts of unstructured data. As S3 is the most common object storage on the market, it is common for many object storage vendors to provide S3-compatible APIs. R2 also provides full S3 compatibility, so existing tools and applications that support S3 will work seamlessly.

A common practice across cloud vendors when it comes to data storage and retrieval is the way they bill the incoming and outgoing traffic, though when most vendors send data into the cloud storage (ingress traffic), this is free or minimally charged. However, the retrievals are usually charged at a much higher rate, and this forces users to use the data from within the cloud vendor's services rather than external services. These exorbitant egress fees pretty much limit the mobility of customer data and thus limit the options available to the customers and developers to build better solutions.

One of Cloudflare's goals, which the company has publicly declared many times, is that they want to slash and even nullify the egress costs. Out of the three dimensions in which object storage is charged – storage, bandwidth, and operations – egress bandwidth is usually the biggest line item in the bill. This charge can often be unpredictable as egress traffic largely depends on customer behavior. The Cloudflare promise is that R2 will not charge for egress traffic at all and, at the same time, the charges for storage and operations will be on par or even lower compared to many other vendors.

In terms of reliability, R2 is built for 99.999999999% (11 9's) of uptime per year. This SLA means the chances of losing the data are extremely low. Like how AWS S3 and AWS Lambda work together (or their counterpart cloud services), R2 is integrated with Workers. This enables the customer to use Workers to transform objects stored in R2 buckets.

Overall, R2 promises the same features as S3, but with better reliability and fewer charges. It is currently in development and there is a waitlist to gain access to it.

Durable objects

Durable objects (**DOs**) are a way of providing strongly consistent storage at the edge (unlike the KV store). It allows you to store data and coordinate across multiple Workers. A DO is defined by a JavaScript class and an ID is given to the objects spawned from the class. There will only be one instance of the DO class with that specific ID across Cloudflare. Multiple Workers can make requests to this object ID and all those requests – probably across the globe – will be routed to the same instance.

DOs are regular JavaScript objects. It also provides a storage API that can persist the object state to disk. Similar to Workers, DOs also implement a fetch method to listen to an HTTP request. Workers can pass information to DOs using request headers or a body, method, or path. Along with DOs, Cloudflare also enabled WebSocket support. This, in combination with DOs, makes it easy to run real-time applications in serverless models.

While Workers are available to try in the free tier, DOs need a subscription/paid plan for you to experiment with them. WebSocket support remains in the early-access phase, free of cost.

Workers and KV – learning by example

Before we start learning how to use Workers, first, you need to have an account with Cloudflare, and then sign up for Cloudflare Workers. Workers has a free plan that should be enough for you to play around with different workers. When you choose your plan and set up your account for Workers, you will also need to have a domain that your workers can be published to. You can either choose to have your own domain, which you can add to your Cloudflare account and manage, or sign up for a free subdomain under the `worker.dev` domain. For this chapter, I am going to use the `safeer.workers.dev` subdomain, which I have signed up for already.

If you want to test how to write Worker code without setting up your account or the necessary tooling, you can try out the Cloudflare playground. The Cloudflare playground is a web-based UI that is available at `https://cloudflareworkers.com`. This UI helps you write Worker code and test it with additional input, header modifications, and more, all from within the browser. While this is a decent way to start experiencing what Workers can do, this won't suffice for our use cases. Here, we will set up development tools in the standard way and experiment with Workers.

Setting up the development environment with Wrangler

Cloudflare Workers has a CLI to manage its configurations and deployments. This utility is called **Wrangler**, a node package that can be set up using npm. Let's look at the setup I followed to set up Wrangler on an Ubuntu machine.

We need to install npm before we can install Wrangler. We will use **Node Version Manager** (**NVM**) to manage node installation. NVM is a bash script-based version manager that can handle multiple versions of node.

NVM installation is covered here: `https://github.com/nvm-sh/nvm`.

Install NVM and then use it to install node and npm:

```
root@serverless101:~# curl -s -o- https://raw.githubusercontent.com/
nvm-sh/nvm/v0.35.3/install.sh | bash
root@serverless101:~# nvm install node
root@serverless101:~#
root@serverless101:~# which node
/root/.nvm/versions/node/v17.3.0/bin/node
root@serverless101:~#
root@serverless101:~# which npm
/root/.nvm/versions/node/v17.3.0/bin/npm
```

Now that npm is available, let us install the worker CLI, Wrangler:

```
root@serverless101:~# npm install -g @cloudflare/wrangler
root@serverless101:~# which wrangler
/root/.nvm/versions/node/v17.3.0/bin/wrangler
```

Now we need to log in to our account using Wrangler. Wrangler will do a one-time authentication with Cloudflare so that it can be used to manage Workers from the command line:

```
root@serverless101:~# wrangler login
Allow Wrangler to open a page in your browser? [y/n]
y
Open a link in your default browser: https://dash.cloudflare.com/
oauth2/auth?response_type=code&client_id=xxxxxx…..
```

If you are on a local machine with a GUI, the link will open in your default browser. If you are on a remote machine, you must copy the link and open it on your local machine. This will take you to the login page – if you are not logged in already – and then complete the OAuth2 authorization process.

If you run into any problems, as an alternative, you can generate an API key by visiting `https://dash.cloudflare.com/profile/api-tokens`. Create an API based on the **Edit Cloudflare Workers** API token template. Once you obtain the API key, you can run the `wrangler config` command to input your API key and complete the CLI configuration.

You can always verify your configuration by running the `wrangler whoami` command, which should show you the account ID and email:

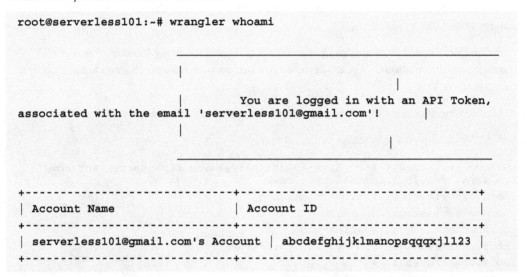

```
root@serverless101:~# wrangler whoami

                        |
                        |                    |
                        |          You are logged in with an API Token,
associated with the email 'serverless101@gmail.com'!          |
                        |
                        |                         |


+------------------------------------+------------------------------------+
| Account Name                       | Account ID                         |
+------------------------------------+------------------------------------+
| serverless101@gmail.com's Account  | abcdefghijklmanopsqqqxjl123        |
+------------------------------------+------------------------------------+
```

Alternatively, you can use environment variables to provide credentials for Wrangler. You could use a combination of `CF_ACCOUNT_ID` and `CF_API_TOKEN` or `CF_EMAIL` and `CF_API_KEY`. This is ideal for CI/CD environments.

Once you have finished setting up Wrangler, you can start creating a simple serverless function in JavaScript.

Creating your first worker

A worker development environment is called a **project**, which acts as a directory with certain minimum required files in various formats. While you can hand-create this directory, Wrangler provides a `generate` subcommand to make your life easier. The command's syntax is as follows:

```
wrangler generate <worker-project-name> <worker-template-url>
```

`worker-project-name`, as its name suggests, is the name of your worker, and when this command is run, Wrangler will create a project directory in your current working directory by this name. This directory will contain some predefined files that will make project creation easier. The last argument in the command – `template URL` – is optional. It allows the user to provide the link to a starter template for Cloudflare Workers and contains the standard project directory structure and reusable code snippets in the required project files. If you omit the template URL, Wrangler will use the default template – `https://github.com/cloudflare/worker-template/` – which is the worker's JavaScript version of the `Hello World` program.

Now, let us create the Hello World project with the name `hello-worker`:

```
root@serverless101:~# wrangler generate hello-worker
 Creating project called `hello-worker`...
 Done! New project created /root/hello-worker
root@serverless101:~#
root@serverless101:~# cd hello-worker/
root@serverless101:~/hello-worker# tree .
.
├── CODE_OF_CONDUCT.md
├── LICENSE_APACHE
├── LICENSE_MIT
├── README.md
├── index.js
├── package.json
└── wrangler.toml
0 directories, 7 files
```

The files in capital letters are not very relevant and are mostly self-explanatory just by looking at their names. However, the other three files are important:

- `index.js`: This contains the Worker code and is the entry point to your Worker.

- `package.json`: This file holds metadata about the project and is a Node.js convention/standard.

- `wrangler.toml`: This is the Wrangler configuration file that carries basic information about the worker and the account. Let us look at the contents of this file:

    ```
    root@serverless101:~/hello-worker# cat wrangler.toml
    name = "hello-worker"
    type = "javascript"

    account_id = ""
    workers_dev = true
    route = ""
    zone_id = ""
    compatibility_date = "2022-01-25"
    ```

Let us look at the significance of `zone_id` and `account_id`:

- `zone_id` is required if you want to deploy to a custom domain. In the absence of this (`zone_id`), the deployment will be done under the `*.workers.dev` subdomain (which is indicated by the `workers_dev = true` configuration).

- `account_id` is the account associated with the zone, if you're using one.

Both `zone_id` and `account_id` can be loaded from the `CF_ZONE_ID` and `CF_ACCOUNT_ID` environment variables, respectively.

Two other relevant configurations keys are as follows:

- `vars`: This defines a list of environment variables that will be accessible from within the worker as global variables. This is a good way to pass configurations to workers.

- `kv_namespaces`: This configures which namespaces will be used by the workers.

Workers can also be deployed to different staging and production environments using the subsection name, `[env.<env-name>]`. The configuration follows the hierarchy and can either inherit or override config values.

To know more about the Wrangler configuration, check out the documentation at `https://developers.cloudflare.com/workers/cli-wrangler/configuration`.

Now that we understand the project layout, let us look at the anatomy of a simple JavaScript Worker.

Workers work in an event-response model. The container runtime will be listening for incoming requests from a client to a specific worker. When that request is received, the container runtime wraps the request (and a whole bunch of context and environment data) into a `fetch` event, spins up a V8 isolate with your worker code, and passes the event to it. When a worker receives an event, it should have a minimum of two pieces of code – a listener and a handler.

Since workers need to listen to a fetch event, the first piece of code is a listener that listens for incoming `fetch` events. The listener will receive the event and use the `respondWith` method of the fetch event to invoke the function that will handle how to respond to the request. This function is the request handler and will accept a request as the argument and return a `Response` object. This will then be passed back to the caller. Now, let us look at the `hello world` example Worker we just created:

```
root@serverless101:~/hello-worker# cat index.js
addEventListener('fetch', event => {
  event.respondWith(handleRequest(event.request))
})
/**
 * Respond with hello worker text
 * @param {Request} request
 */
async function handleRequest(request) {
```

```
    return new Response('Hello worker!', {
      headers: { 'content-type': 'text/plain' },
    })
  }
```

The code is pretty simple and self-explanatory, so I will let you figure this out. Now, let us take a quick peek at some of the methods and object syntaxes.

addEventListener

This function registers triggers that a worker script can execute. The syntax is as follows:

```
addEventListener(type, listener)
```

The first argument is the type of event – two types are supported currently: fetch and scheduled. fetch is for receiving web requests, while scheduled is for the scheduled execution of workers, such as a cron job. You can add both fetch and scheduled event listeners in the same worker script. The second argument is the listener function, which is used to handle an incoming event. It takes the event (fetch or scheduled) as its argument.

Request

Request is an object that represents the incoming HTTP request and is a part of FetchEvent, which will be passed to the listener. The object has a bunch of properties that would be expected from an HTTP request, such as a body, headers, method, and more.

Response

Response is an object that contains the response to the caller. It takes two arguments: the body of the response (optional) and an options object that contains a set of HTTP properties that can be applied to the request. These include status (HTTP status code), statusText (HTTP status message) and a headers object that contains a list of headers.

You can learn more about these objects, methods, and much more here: https://developers. cloudflare.com/workers/runtime-apis.

Deploying your worker

Now that we have created the first worker and understood its structure, let us deploy it to Cloudflare. We will use the `wrangler dev` command to run a local instance of the worker and test it before deploying it to Cloudflare. The dev instance listens on port 8787 on localhost. Since I am running this experiment on a remote host, I want it to listen on the public IP. For this, I will use the `-i` argument with the `wrangler dev` command. To see all the supported parameters, run `wrangler dev –help`:

```
root@serverless101:~/hello-worker# wrangler dev -i 0.0.0.0
Listening on http://0.0.0.0:8787
watching "./"
```

Now, I can access this URL from my local machine:

```
$ curl http://142.93.221.66:8787/
Hello worker!
```

As you can see, our `hello-world` (worker!) program is up and running in dev. Now, let us deploy it to Cloudflare. Press *Ctrl* + *C* to abort the dev instance of the worker and return to the shell. The `wrangler publish` command will build, upload, and deploy your code to my domain – that is, `safeer.workers.dev`. The URL format for any project is `<project-name>.<cloudflare-acccount-free-subdomain-name>.workders.dev`:

```
root@serverless101:~/hello-worker# wrangler publish
Basic JavaScript project found. Skipping unnecessary build!
Successfully published your script to
  https://hello-worker.safeer.workers.dev
root@serverless101:~/hello-worker# curl  https://hello-worker.safeer.
workers.dev
Hello worker!
```

As you can see, the Worker has been deployed to the domain and is now available on the internet.

Now that the first Worker is up and running, let us explore how to use the Workers KV store.

Workers KV store

As discussed earlier, KV is a globally distributed, eventually consistent, and low-latency key-value data store. To use KV with your worker, you need to create a KV "namespace" and bind it to your worker. You can consider a KV namespace as a database within KV that allows you to separate the set of keys you want for a particular purpose. Each namespace, when created, will have the name that you chose for it and an ID that is assigned by Cloudflare. Within a Worker, the binding is the global variable that exposes the get and set methods for the namespace. The binding is configured by using a name and the Cloudflare-assigned ID of the namespace. Let us learn how to create a namespace and how to bind it in a Worker configuration.

The following code shows how to create a KV binding named `hello_kv`. Note that a binding can only contain alphanumeric and _ characters, and cannot begin with a number. The following command must be run from the root directory of your project:

```
root@serverless101:~/hello-worker# wrangler kv:namespace create
"hello_kv"
Creating namespace with title "hello-worker-hello_kv"
Success!
Add the following to your configuration file:
kv_namespaces = [
     { binding = "hello_kv", id = "f7f525e769ec44b8817d93d50d6173dc" }
]
```

Note that the `name` namespace is in the `<worker-project-name>-<binding>` format – in this case, this will be `hello-worker-hello_kv`. I will append the last three lines of the output of the `wrangler.toml` file.

KV supports three methods to manipulate data: a `put` method to write key-value pairs to the KV store, a `get` method to retrieve the value by key, and a `delete` method to delete the value using the key. You can invoke these methods from within your code by using the Wrangler CLI or using the Cloudflare API. Note that in Cloudflare, writing to the same key is limited to one per second. This is in line with the KV design principle to support a *write-once-read-many* nature. So, if you are planning to implement some sort of a timer or high-volume update data store, KV is not a good choice.

The following commands will list your namespaces and then create, get, and delete a key-value pair:

```
root@serverless101:~/hello-worker# wrangler kv:namespace list
[{"id":"f7f525e769ec44b8817d93d50d6173dc","title":"hello-worker-hello_
kv"}]


root@serverless101:~/hello-worker# wrangler kv:key put MyKey MyValue
-b hello_kv
Success


root@serverless101:~/hello-worker# wrangler kv:key get MyKey -b hello_
kv
MyValue


root@serverless101:~/hello-worker# wrangler kv:key delete -f MyKey -b
hello_kv
Deleting key "MyKey"
Success
```

To demonstrate the KV use case, I am going to create a simple API that will fetch the current market price of cryptocurrencies from the Coinbase public API. We will limit the type of cryptocurrency and the real/traditional currency pairs support, called Currency Pair. For example, ETH-USD means Etherium price in US Dollars. We will also cache the result for a minute so that we don't hit the Coinbase API too often. This means that if multiple requests are made to our API for the same currency pair within a minute, they will see the cached price rather than the spot price of that second (crypto traders won't like this API!). We will pass the currency pair as part of the path.

An example `curl` call to the API for Etherium in USD is as follows:

```
$ curl -s https://api.coinbase.com/v2/prices/ETH-USD/spot|jq .
{
  "data": {
    "base": "ETH",
    "currency": "USD",
    "amount": "2610.57"
  }
}
```

Let us create a fresh Worker project using Wrangler:

```
root@serverless101:~# wrangler generate coinbase-spot
 Creating project called `coinbase-spot`...
 Done! New project created /root/coinbase-spot
root@serverless101:~# cd coinbase-spot/
```

We will create two KV namespaces – one for caching the result for 1 minute and another for keeping the whitelist of allowed currency pairs. In the currency pair namespace, each currency pair code, such as ETH-USD, will be the key; the value would be 1. The value doesn't matter since all we are going to check is if the key exists in the namespace. For the cache, we will use another namespace and a feature built into the KV store. For each KV pair, you can add an expiry/**time-to-live** (**TTL**) and some metadata. We will set a TTL of 60 seconds for each item we add to the cache namespace. While cache namespace entries will be managed from within the code, the whitelist will be managed from outside using the Wrangler CLI:

```
root@serverless101:~/coinbase-spot# wrangler kv:namespace create "cb_
whitelist" --preview
Creating namespace with title "coinbase-spot-cb_whitelist_preview"
 Success!
Add the following to your configuration file in your kv_namespaces
array:
{ binding = "cb_whitelist", preview_id =
"0894499ebabf4f1baa6967c608c1e668" }
root@serverless101:~/coinbase-spot# wrangler kv:namespace list
[{"id":"0894499ebabf4f1baa6967c608c1e668","title":"coinbase-spot-cb_
```

```
whitelist_preview"},{"id":"1cd5012fb9d349399dffc0e241dd23be","title":"
coinbase-spot-price_cache_preview"},{"id":"35da953682d845bebebbe02cb32
e47f2","title":"coinbase-spot-whitelist_preview"}]
root@serverless101:~/coinbase-spot# wrangler kv:namespace create "cb_
price_cache" --preview
Creating namespace with title "coinbase-spot-cb_price_cache_preview"
 Success!
Add the following to your configuration file in your kv_namespaces
array:
{ binding = "cb_price_cache", preview_id =
"5eb7ad4fcd094052bc9916abbbf60b8e" }
```

```
## Add the kv_namespaces to your wrangler.toml and verify
```

```
root@serverless101:~/coinbase-spot# toml get --toml-path wrangler.
toml  kv_namespaces
[{'binding': 'cb_whitelist', 'preview_
id': '0894499ebabf4f1baa6967c608c1e668', 'id':
'0894499ebabf4f1baa6967c608c1e668'}, {'binding': 'cb_price_
cache', 'preview_id': '5eb7ad4fcd094052bc9916abbbf60b8e', 'id':
'5eb7ad4fcd094052bc9916abbbf60b8e'}]
```

```
## Add the whitelist values using wrangler
```

```
root@serverless101:~/coinbase-spot# for pair in BTC-USD BTC-INR ETH-
USD ETH-INR DOGE-USD DOGE-INR;do  wrangler kv:key put ${pair} "1" -b
cb_whitelist;done
 Success
 Success
 Success
 Success
 Success
 Success
```

Open the index.js file and add the following content. The code is mostly self-explanatory and simple.
I have not added all the corner cases and exception handling to keep the code short for this book:

```
root@serverless101:~/coinbase-spot# cat index.js
addEventListener('fetch', event => {
  event.respondWith(handleRequest(event.request))
})
```

```
async function handleRequest(request) {
  const path = new URL(request.url).pathname.split('/')
  pair_pattern=/^[A-Z]+\-[A-Z]+$/
  if( path.length !== 3 || path[1] !== 'prices' || pair_pattern.
test(path[2]) !== true  ){
    console.log("Unsupported API call")
    return new Response( '{"message":"ERROR: Unsupported API call"}',
{
        headers: { 'content-type': 'text/json' },
        status: 400,
        satusText: 'Bad request',
         })
  }
  const currency_pair = path[2]
  const wl_value = await cb_whitelist.get(currency_pair)
  if( wl_value == null ){
    return new Response( '{"message":"ERROR: Currency pair not
allowed"}', {
        headers: { 'content-type': 'text/json' },
        status: 400,
        satusText: 'Bad request',
      })
  }
  let prices = null
  pair_cache = await cb_price_cache.get(currency_pair)
  if( pair_cache !== null ){
    prices = JSON.parse(pair_cache)
    console.log("pair taken from cache")
  }
  else {
    const base_url = 'https://api.coinbase.com/v2/prices/'
    const url = base_url.concat(currency_pair,'/spot')
    const result = await fetch(url)
    prices = await result.json()
    console.log('pair fetched from Coinbase' )
    await cb_price_cache.put(currency_pair,JSON.
stringify(prices),{"expirationTtl": 60})
  }
  return new Response(JSON.stringify(prices['data']), {
    headers: { 'content-type': 'text/json' },
  })
}
```

Publish the project:

```
root@serverless101:~/coinbase-spot# wrangler publish
 Basic JavaScript project found. Skipping unnecessary build!
 Successfully published your script to
 https://coinbase-spot.safeer.workers.dev
```

Now, let us test the project using the DOGE-USD currency pair. First, it should retrieve data from the Coinbase API and then from the cache (within 1 minute of the first request):

```
$ curl -w "\nTime taken to fetch data in seconds: %{time_total}\n" -s
https://coinbase-spot.safeer.workers.dev/prices/DOGE-USD
{"base":"DOGE","currency":"USD","amount":"0.1443"}
Time taken to fetch data in seconds: 2.311952


$ sleep 20;curl -w "\nTime taken to fetch data in seconds: %{time_
total}\n" -s https://coinbase-spot.safeer.workers.dev/prices/DOGE-USD
{"base":"DOGE","currency":"USD","amount":"0.1443"}
Time taken to fetch data in seconds: 0.523443
```

As you can see, the first call (which goes to the Coinbase API) takes roughly 2.31 seconds, whereas the result fetched from KV only takes about 0.52 seconds.

You may have noticed that there are a couple of `console.log` lines in the script. This is to demonstrate the debugging capability of workers by fetching logs from Cloudflare. The way to retrieve this (other than using the Cloudflare UI) is to run the `wrangler tail` command against the worker. The result would be a stream of JSON objects that you can parse or filter using tools such as jq. I am going to start the log fetching against the `coinbase-spot` worker and then send a couple of web requests for DOGE-INR one after the other. Let us see what shows up in the log:

```
root@serverless101:~/coinbase-spot# wrangler tail coinbase-spot|jq
.logs
 Connected! Streaming logs from coinbase-spot... (ctrl-c to quit)

[
  {
    "message": [
      "pair fetched from Coinbase"
    ],
    "level": "log",
    "timestamp": 1643295005914
  }
```

```
]
[
  {
    "message": [
      "pair taken from cache"
    ],
    "level": "log",
    "timestamp": 1643295020066
  }
]


^C
```

As you can see, my two consecutive requests had two different log lines showing that the first request was fetched from Coinbase and that the second one was fetched from the cache. This is as expected.

Now that we have seen how to create and run a JavaScript-based worker, let us explore the language support by creating a simple Python hello-world worker. We are not going to create any fresh code; instead, we are going to use a Worker template provided by Cloudflare. The idea is to learn how to set up and run a Python Worker.

Python worker development requires python3, the virtualenv tool, and the pip3 installer. In addition, we need TypeScript, an OOL that can precompile a large part of Python code into JavaScript. This essentially helps Python workers run on Isolates. You already have Wrangler and npm installed, so that is all you require to create a Python Worker:

```
root@serverless101:~#wrangler generate python-worker https://github.
com/cloudflare/python-worker-hello-world
root@serverless101:~#cd python-worker
root@serverless101:~/python-worker#python3 -m virtualenv venv
root@serverless101:~/python-worker# source venv/bin/activate
(venv) root@serverless101:~/python-worker# pip3 install transcrypt
(venv) root@serverless101:~/python-worker# wrangler publish
<<output truncated>>
Built successfully, built project size is 5 KiB.
Successfully published your script to
https://python-worker.safeer.workers.dev
(venv) root@serverless101:~/python-worker# curl -s  https://python-
worker.safeer.workers.dev
Python Worker hello world!
```

For more detailed instructions, check out https://github.com/cloudflare/python-worker-hello-world.

Now that we know how the worker developer ecosystem works, let us look at some of the other Worker features:

- **Environment variables**: Environment variables are treated like other bindings and are available to the Worker as global variables. You can add environment variables by declaring them in your `wrangler.toml` file. Environment variable values are always plaintext.

- **Secrets**: Secrets are used to keep passwords, API keys, and other secrets required by the worker. They are similar to environment variables and are available to the worker as global variables. The difference is that the secrets are not visible once they are declared. You can create secrets either using the `wrangler secrets put <binding_name>` command or via the dashboard. As secrets are bindings, you need to declare them in the `wrangler.toml` file (without the secret value).

- **Routes**: Routes are used to send traffic coming from a particular domain/path URL pattern to a worker. Routes are only available on custom domains hosted in Cloudflare. They can be configured from the web console, Cloudflare, or the Cloudflare API. The Wrangler CLI only allows you to list and delete routes.

- **Cron**: crons are triggers that can map a cron expression to a worker. This feature enables users to run scheduled tasks outside of HTTP endpoint ecosystems, making it more useful as a standalone task executor. crons can only be configured via the Cloudflare dashboard and API, not through Wrangler. A Worker that can be invoked via cron should have a listener configured to listen to `scheduled` events.

- **Worker Sites**: Worker sites allow static web content to be deployed alongside a worker. Cloudflare Pages came after worker sites and is the preferred way to host static content nowadays.

- **Cache API**: This allows Workers access to the Cloudflare Edge Cache. You can put content into the cache as well as retrieve it easily. Please note that the content of the cache is stored locally in the data center where the request originated.

Now that we have had some exposure to developing and configuring workers, let us look at our project.

Project

For this chapter's project, we are going to build a GeoIP API. GeoIP services provide location information about a particular IP. Usually, the amount of information you receive varies from provider to provider. Most of them provide a minimum of longitude and latitude, as well as the country and city where the IP is located. GeoIP data is not always fully accurate, but it can always provide reasonable location information. A lot of mobile and web applications use GeoIP services to know the location of their customers. This helps them personalize the experience, show them a location-specific web page, or use it for legal or auditing purposes.

There are a lot of GeoIP service providers. Some provide it as REST APIs, while others provide downloadable databases and SDKs, along with rolling updates. So, why do we need a new GeoAPI of our own? The APIs and details of data provided vary from provider to provider. There are request limits on how many queries you can make to these endpoints, on free as well as most paid services. So, our idea is to create a middleman API that provides a uniform GeoIP service to our applications/clients while we manage our access to the provider – including the limits and authentication mechanisms behind the scenes. Besides this, you can add a caching layer that will reduce the load on the origin API servers. If you want to extend this service further, you can also add multiple providers to the API and balance the requests between them.

For this exercise, we are going to use the Maxmind GeoLite2 API. Maxmind is a top geolocation provider – it offers both APIs and the ability to download databases offline. Their free version of GeoIP is not as accurate as the paid one, but for our demonstration, this will do just fine. To start, you need to create a free account with them and then create a license key that will be needed to authenticate with their API endpoint. More details on Maxmind can be found at `https://dev.maxmind.com/geoip/geolite2-free-geolocation-data` and `https://blog.maxmind.com/2020/12/geolite2-web-service-free-ip-geolocation-api`.

Implementation

We will create a simple API endpoint that will take an IP address as the argument and return its latitude, longitude, and country. We will implement authentication for our API using simple API keys (stored in KV) and implement some basic caching mechanisms.

The API keys will be added to a KV namespace using Wrangler (since building an API key management platform is outside the scope of this project). We will also need a KV namespace to store the API credentials we will use to call the Maxmind API. For caching, we could use either KV or the Cloudflare cache. Since we looked at caching using KV in the Coinbase example, we will use the Cloudflare cache in this project. Please note that, unlike KV, the Cloudflare cache is not global and is local to the DC that the web request came through.

For detailed instructions and code, please visit the GitHub repository of this book: `https://github.com/PacktPublishing/Architecting-Cloud-Native-Serverless-Solutions/tree/main/chapter-6`

More vendors in edge computing and JAMStack

Outside the cloud vendor landscape, there are CDNs and other dedicated vendors who offer serverless at the edge. The following are some examples:

- Akamai Edge
- IBM Edge Functions

- Netlify functions
- Vercel Edge Functions
- Gatsby Cloud Functions

There are many more providers, but these top providers are worth checking out. Several providers support JAMStack hosting, some of which are as follows:

- Netlify
- Vercel
- GitHub/GitLab Pages
- Azure Static WebApps
- Gatsby Cloud

There are many more established and up-and-coming serverless providers. While this list is not comprehensive, it will give you an idea of the landscape.

Cloudflare has been an established player. Its foray into edge functions and the serverless world is exciting as they have world-class expertise in running internet infrastructure at scale and with efficiency. Functions, Pages, and KV have already been well established and widely adopted. If the release of durable objects and R2 is any indication, we can expect more powerful serverless offerings at the edge soon.

Summary

In this chapter, we looked at a vendor that is much different than the traditional cloud providers. As you can see, Cloudflare capitalized on its existing infrastructure to bring serverless to the edge, along with supporting services such as KV, Pages, and others. We looked at the key serverless technologies offered by them and understood the lay of the land. We also got our hands dirty by writing some workers and implementing a small project.

Cloudflare has been very innovative in bringing new services to the edge. The services that are in beta and private preview are going to mature soon and will be available to the public. Given their drive for performance, low latency, and other goals that enable customers to implement faster and better business solutions, it is only fair to assume that they would want to take this further with more services. We will soon see a more feature-rich edge computing platform that promotes serverless as its primary development philosophy.

In the next chapter, we will explore Kubernetes and three of the top serverless solutions that are built on top of it.

7

Kubernetes, Knative and OpenFaaS

Docker and Kubernetes have revolutionized the application and infrastructure world. While Docker facilitated the containerization of applications with ease, Kubernetes took it to the next level. Originally developed by Google, Kubernetes is a container orchestration platform. It is open source, widely popular, and highly extensible. The biggest advantage is that it can run across all public and private clouds as well as data centers.

Kubernetes allows running containerized applications at scale and has become a favorite of developers and infrastructure operators alike. With the wider acceptance of serverless technologies, it was inevitable that serverless platforms that work with Kubernetes would be developed and deployed. There are several platforms that can run on top of Kubernetes and provide serverless services. Some of them provide simple FaaS services while others provide a complete event processing system. For this chapter, we have picked two of the top serverless platforms – Knative and OpenFaaS.

The core topics covered in this chapter are as follows:

- The fundamentals of containerization and Docker
- Kubernetes core concepts
- How to run a simple Kubernetes cluster and deploy a Docker containerized application
- Select advanced concepts in Kubernetes
- Knative
- OpenFaaS

Containerization and Docker fundamentals

Before we discuss Kubernetes, we need to have a solid understanding of what role containers and Docker play in it. Containers facilitate the packaging of application code, runtime, configuration, and systems libraries as a single unit of deployment. This allows developers to ship their applications without worrying about dependency packaging, or where the application will run. It avoids the classic problem in software deployment where the developer says, "*It works on my machine,*" but when deployed into a production environment, the dependencies break. A container is a lightweight wrapper around the application that isolates the filesystem, network, and process space from the rest of the machine it is running on. This ensures that the application works as expected across all sorts of systems, from the developer machine, through the test and staging servers, to the production server infrastructure.

Containers are not a new concept; they have been in existence for a long time in one form or another. chroot is a concept that has been around since the 1980s and is used to change the root directory of a Linux/Unix process to restrict the process from accessing any files beyond its root directory. This provided some amount of security in the case where a process is compromised. FreeBSD jails were started in 2000 and were used to sandbox a process, which meant that all activities of the process were limited to the confines of the jail. A container technology that was closest to Docker is LXC, aka Linux containers; in fact, Docker is based on the concepts of LXC and the initial versions directly depended on the LXC runtime.

Let us look at what the fundamental features of a container are:

- A container is a way to group processes and control the system resources that are allocated to them
- All containers in a host share the host kernel
- Containers give the feeling that they are lightweight VMs but they are actually virtual environments for processes to run
- They use an underlying isolation mechanism provided by the Linux kernel to provide a virtual environment
- They have a set of tooling to manage isolation and resource allocation
- There are different OS images for different containers, independent of the host OS version

The following diagram shows how containers are different from virtual machines:

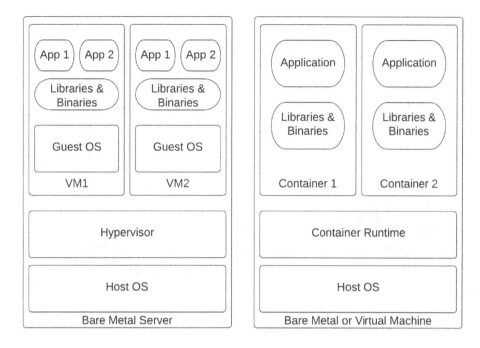

Figure 7.1 – Containers versus virtual machines

While support for containers exists in Linux, the tooling necessary to run them used to vary between solutions. Docker has evolved to provide the following features as part of its container tooling:

- Resource isolation
- Networking
- Storage
- Image building and management
- Union filesystem to support a layered root filesystem
- Docker Engine to manage container lifecycle
- Docker API and CLI for the automation of container lifecycle management

The following diagram shows the architecture of the Docker ecosystem:

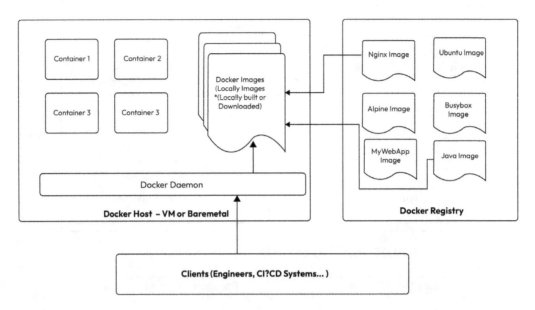

Figure 7.2 – Docker ecosystem

Now that we have a good understanding of the Docker ecosystem, let's delve into a quick Docker primer that will help you understand Docker images.

Docker images

A Docker container can run one process isolated from the host with select software packages and a runtime. To begin with, an existing Docker image should be used as the base, and then additional software packages and tooling can be layered on top of it.

A Docker image is a way to provide an immutable and reproducible virtual environment. This virtual environment would contain the software application, libraries, and binaries, along with the environment variables, ports to expose, and the command to execute when a container is created out of that image. These images are created as read-only and any modifications made on top of an image are not persistent unless you build an image with the changes and use that for your future containers. If you build such an image, it will also remain read-only and would allow you to reproduce the same virtual environment repeatedly.

Docker provides a file template called a Dockerfile, where we can add instructions on how to build an image, including the base image from which we want to create the OS, packages to install, the working directory, and the command to run, among other things. Then, you can use the Docker CLI to build an image out of this file. The images you created can either live on the host where you created them (and can only be used on that host) or can be pushed to a central repository called `docker-registry`. This registry can either be public or private. Docker Inc maintains the most well-known public registry themselves, called Docker Hub. There are other cloud providers and organizations that provide public as well as private registries – `https://docs.docker.com/engine/install/`

> **Note**
>
> Now head to the official Docker "getting started" documentation and familiarize yourself with how to install Docker and manage containers. Refer to `https://docs.docker.com/get-started/`.

Docker offers a standardized way to deploy applications across different servers and platforms in a homogenous way without worrying about dependencies, environments, and configuration differences. But this has also introduced some challenges in deploying such applications on a large scale. In the following section, we will examine the solutions for that.

Container orchestration and Kubernetes

Container technologies have transformed the way organizations build, package, and deploy software. The obvious advantages of dependency management and resource isolation have made it a favorite of developers as well as infrastructure operators. But there is one problem: operational complexity.

Microservices architecture adoption has meant that every reasonably sized business use case is translated into several microservices rather than one or two monoliths. To scale these microservices using containers, multiple containers have to be spawned for each microservices. This means there is increased complexity in managing the lifecycle of these containers, how they communicate via a network, how they load balance, how they access data kept in storage or external datastore, and so on. Soon, any organization with a large number of microservices will start facing operational challenges and scalability issues.

These challenges lead to the invention of a technology stack that can take away all the operational challenges and orchestrate the lifecycle management of containers. That technology is generally referred to as container orchestration. Container orchestration automated away most of the pain associated with manually managing the following operations:

- Provisioning and deployment of containers
- Operations to do with the redundancy and availability of containers
- Scaling up or removing containers to spread application load evenly across host infrastructure
- Movement of containers from one host to another if there is a shortage of resources in a host, or if a host dies

- Allocation of resources between containers

- External exposure of services running in a container with the outside world

- Load balancing and service discovery

- Health monitoring of containers and hosts

- Configuration of an application in relation to the containers running it

Docker, being the leader in containerization, attempted to solve this issue with their own product called Docker Swarm. It came out in 2013 and generated a lot of interest and adoption. In 2014, Google came out with Kubernetes, which was modeled after their internal cluster management and scheduling solution called Borg (which was later renamed Omega). A number of organizations joined the Kubernetes community and active development and community adoption followed soon. In 2015, Google partnered with the Linux Foundation to form the **Cloud Native Computing Foundation (CNCF)**, to build an ecosystem and community around containers and other related technologies. It has since grown into a foundation that curates and maintains a number of technologies that enable the organization to run scalable applications in public and private clouds. Apache Mesos was another contender in the orchestration world, but over the years Kubernetes has surpassed both Mesos and Swarm by leaps and bounds and has emerged as the de facto leader in container orchestration.

The following are some of the features offered by Kubernetes:

- Automation of workflows and tasks that were manual when running containers directly

- Autoscaling of containers based on traffic and other parameters

- Self-healing of containers

- Load balancing and traffic routing

- Secret and configuration management

- Infrastructure-as-Code support

- Bin packing – the efficient utilization of hardware by placing containers on servers where resources are free

- Storage management

- Runs on any cloud, VM, or hardware platform

- Batch jobs by running containers ad hoc

- Automated deployment and rollbacks

Kubernetes helps companies to optimize their infrastructure and improve developer velocity. Deployments are much easier and more controlled. There is a large ecosystem of additional cloud-native services coming up around Kubernetes that help improve its utility and general infrastructure efficiency. In the next section, we will look at the architecture and components of Kubernetes that make it such an efficient infrastructure orchestrator.

Kubernetes architecture and components

Kubernetes works as a cluster; it has one or more masters that control the orchestration and infrastructure management and a number of nodes that actually run the container workloads. The master server is responsible for the control plane, which is constantly monitoring and managing the containers on the worker nodes. Before we examine the control plane, let's understand a few terms/concepts related to Kubernetes. Most of these concepts represent a type of resource within Kubernetes and fall under the category of a Kubernetes object. Let's have a look at them:

- **Pod**: One or more containers bundled together and treated as an application. This is the unit of deployment in Kubernetes and a single instance of the application.

- **Namespcace**: Logical boundary within a Kubernetes cluster that is used to segregate applications belonging to different teams or organizations. It can be considered a virtual cluster within the physical cluster.

- **Volumes**: Similar to Docker volumes, but instead of a single container, a volume is available to the entire Pod in Kubernetes. There is also the concept of persistent volume which will point to storage provided by external storage systems.

- **ReplicaSet**: An object that ensures a specified number of Pods of an application are running at a given time.

- **Deployment**: An object for declaratively updating the state of an application. It is applicable to both ReplicaSets and Pods.

- **DaemonSets**: An object that ensures all or a selected set of nodes run a copy of a Pod. This is a good use case where a monitoring or security agent has to be run on all nodes.

Now that we have an understanding of some of the basic concepts of Kubernetes, let us get back to the control plane. The following are the components of the control plane that are running as services within the masters:

- **kube-apiserver**: The service that exposes the Kubernetes API. This REST API is the interface to the Kubernetes control plane.

- **kube-scheduler**: This service keeps watch for Pods that are not allotted a node, looks for a node that has sufficient resources for the Pod, and places it on that free node.

- **kube-controller-manager**: There are a number of controllers in the Kubernetes control plane. Their job is to keep an eye on specific objects within the Kubernetes cluster and take action based on their state changes. For example, a node controller would keep an eye on all nodes and take action when one goes down. The controller manager, as you can tell from the name, is a daemon that manages all the controllers.

- **etcd**: Highly available key-value store that stores all the cluster data.

- **cloud-control-manager**: An optional component that is used to link up to your cloud provider's API, if you are running the Kubernetes cluster in the cloud.

Those are the components that constitute the control plane and run on the masters. Now let us look at the services that run on the nodes:

- **kubelet**: The agent that runs on all nodes, communicates with the control plane, and executes the control plane's instructions to manage the container lifecycles.

- **kube-proxy**: A network proxy that handles traffic between Pods as well as the network traffic router. It manages load balancing, routing, subnetting, and so on.

- **Container runtime**: This is the software component on the nodes that is responsible for managing the containers. There are multiple container runtimes, including containerd, CRIO, and many more. The most prominent one is still Docker Engine, and it is used by most organizations.

The following diagram depicts the Kubernetes cluster architecture:

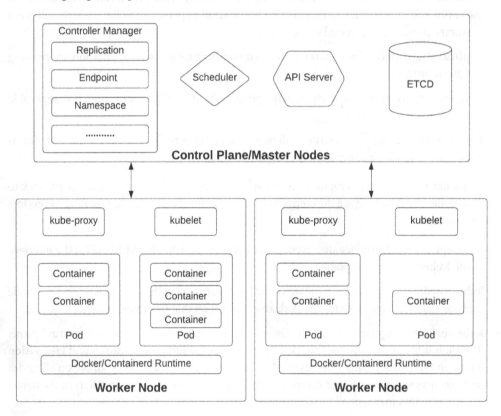

Figure 7.3 – Kubernetes architecture

Now that we have a basic understanding of the Kubernetes architecture, let us have a quick *how-to* to set up a developer Kubernetes cluster.

Kubernetes how-to with minikube

Before we delve into the *how-to* part of things, let's have a look at what minikube is.

While Kubernetes is a fantastic project, developers will find it difficult to maintain multi-node clusters for their developer environment. For this reason, Kubernetes maintains a companion project called minikube. minikube can work with a container runtime such as Docker to simulate a full-fledged Kubernetes cluster on a single machine.

We will be setting up Docker and minikube on an Ubuntu machine. For installation on different platforms, refer to the relevant software documentation. Let us start with the installation of both components. I will be trimming down most of the output and only showing the most relevant output for better readability.

Please note that the minimum requirement for running minikube are as follows:

- Two or more CPUs
- 2 GB of free memory
- 20 GB of free disk space
- A virtual machine or container runtime such as KVM, Docker, or VirtualBox
- An active internet connection

Following are the steps for the installation:

1. Make sure you are not a root user and are using `sudo` instead:

    ```
    safeer@serverless101:~$ sudo apt install docker.io
    ```

2. Ensure you are part of the `sudo` and Docker groups:

    ```
    safeer@serverless101:~$ groups
    safeer sudo docker
    ```

3. Install minikube. Refer to `https://minikube.sigs.k8s.io/docs/start/`:

    ```
    safeer@serverless101:~$ curl -LO https://storage.googleapis.com/
    minikube/releases/latest/minikube_latest_amd64.deb

    safeer@serverless101:~$ sudo dpkg -i minikube_latest_amd64.deb
    ```

4. Start the minikube cluster. This step will fail if you don't have sufficient CPU/RAM/disk resources:

```
safeer@serverless101:~$ minikube start
```

5. Set up an alias for the `kubectl` command (which is the CLI used for managing Kubernetes):

```
safeer@serverless101:~$ echo "alias kubectl='minikube kubectl'"
>> ~/.bashrc
```

6. Verify that Kubernetes is running:

```
safeer@serverless101:~$ kubectl cluster-info
Kubernetes control plane is running at https://192.168.49.2:8443
CoreDNS is running at https://192.168.49.2:8443/api/v1/
namespaces/kube-system/services/kube-dns:dns/proxy
```

7. This gives us a working, single-node Kubernetes cluster. Now let us deploy a sample application. We will be creating a "hello world" app to show how a sample application is deployed and accessed in Kubernetes. We are going to deploy a Pod (a group of one or more related application containers) using a Deployment, which is responsible for health checks and monitoring the uptime of the Pod containers. After that, we will expose the service as a Kubernetes Deployment and access the endpoint. Let's start with creating a Deployment from a sample app. List its Deployment and Pod:

```
safeer@serverless101:~$ kubectl create deployment hello-node
--image=k8s.gcr.io/echoserver:1.4
deployment.apps/hello-node created
safeer@serverless101:~$ kubectl get deployments
NAME        READY   UP-TO-DATE   AVAILABLE   AGE
hello-node  1/1     1            1           88s
safeer@serverless101:~$ kubectl get pods
NAME                         READY   STATUS    RESTARTS   AGE
hello-node-6b89d599b9-tdfwg  1/1     Running   0          98s
```

8. Now let us create a load balancer and get its URL:

```
safeer@serverless101:~$ kubectl expose deployment hello-node
--type=LoadBalancer --port=8080
service/hello-node exposed

safeer@serverless101:~$ kubectl get services
NAME        TYPE          CLUSTER-IP     EXTERNAL-IP   PORT(S) AGE
hello-node  LoadBalancer  10.100.90.249  <pending>
8080:32138/TCP    18s
kubernetes  ClusterIP     10.96.0.1      <none>        443/
TCP         119m
```

9. In a cloud provider environment, the external IP will show a load balancer IP. For minikube, we need to use the following command to get a working URL to the app:

```
safeer@serverless101:~$ minikube service hello-node --url
http://192.168.49.2:32138
```

10. Access the deployed app:

```
safeer@serverless101:~$ curl http://192.168.49.2:32138
CLIENT VALUES:
client_address=172.17.0.1
...OUTPUT TRUNCATED...
BODY:
-no body in request-
```

11. Clean up the resources created:

```
safeer@serverless101:~$ kubectl delete service hello-node
service "hello-node" deleted
safeer@serverless101:~$ kubectl delete deployment hello-node
deployment.apps "hello-node" deleted
```

12. If you want to restart from a clean slate, you can run the following command to remove the minikube VM:

```
minikube stop
minikube delete
```

This concludes our quick primer on minikube. Now that you know how to run a Kubernetes cluster on your developer machine, let's learn about the first serverless framework in our list – Knative.

Knative

Knative is an open source project that provides a set of components and objects for Kubernetes to run stateless as well as event-processing workloads. It was created by a consortium of companies, including Google, IBM, Red Hat, and many more, who felt the need to have a framework like this that can work across Kubernetes installations in various cloud and on-premises infrastructures. The motivation to do so was the same as the motivation and demand for serverless platforms. Developers want to be able to run their code and not have any part in managing the infrastructure. They should be free to run any application written in any language and framework they choose. This is why cloud vendors started providing their own flavors of serverless. But they also had to introduce reasonable limits and compatibility restrictions in order to run an effective public cloud offering. Knative wants to provide more freedom to developers while taking away the complexity. In the following sections, we will learn about the components of Knative and how they are used.

Knative components

Knative fundamentally has two primitives or building blocks: Knative Serving and Knative Eventing. The primitives are implemented as two independent projects and can be deployed separately or together. Let us look at each of them.

Knative Serving

Knative Serving, as the name suggests, is about serving traffic. This entails the following features:

- Deploying stateless containers fast and efficiently

- Autoscaling of containers – from zero to N

- Networking and traffic management

- Code and configuration deployments as a unit; a point-in-time snapshot

These features are implemented as four Kubernetes objects/resources. Let us take a look at them:

- **Service**: This object manages the entire lifecycle of a Knative application. It is also responsible for automatically creating the other resources required by the service resource.

- **Route**: Maps the URL to a version of your serverless application. There are different methods by which you can map the request to one or more revisions. By default, the route maps to the latest revision, or you could have scenarios such as splitting the traffic percentage-wise and sending it to the different revisions deployed.

- **Configuration**: This resource maintains the desired state of the serverless application. It keeps the configuration and the code bifurcated and is responsible for maintaining revisions. Any modification to the existing configuration creates a new revision.

- **Revisions**: A revision resource is a point-in-time snapshot of the application code and associated configuration. Every modification to either the code or config creates a new revision. Revisions can't be modified and any number of past revisions can be maintained as required. Autoscaling essentially scales deployments of active revisions as desired. If a revision is not mapped with a route, it could get garbage collected and all related resources could be deleted.

The Knative Serving resource hierarchy and interconnections are as follows:

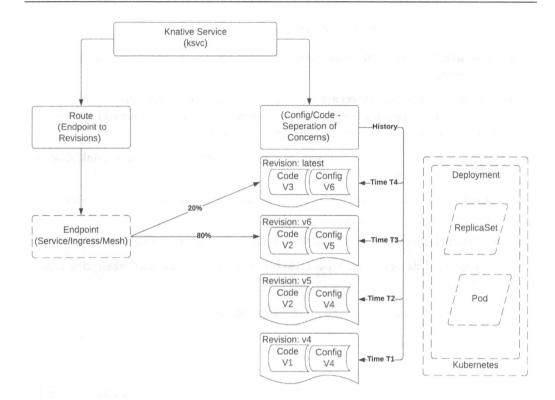

Figure 7.4 – Knative Serving

Next, we will look at Knative Eventing.

Knative Eventing

Knative Eventing is a set of tools for routing events from a producer to consumers, providing developers with a way to integrate event-driven architecture in their applications. It allows declaratively binding event producers and Knative services. Knative-supported events are in the format specified by CloudEvents (https://cloudevents.io/). It defines a set of metadata attributes that should be associated with each event making it easy to process. It uses the HTTP POST method to send and receive events between sources and sinks.

Knative allows a loosely coupled service architecture where producers and consumers can work independently. The Knative official documentation articulates it in this way:

> *A producer can generate events before a consumer is listening, and a consumer can express an interest in an event or class of events that is not yet being produced. This allows event-driven applications to evolve over time without needing to closely coordinate changes.*

Let us looks at the core components of the Eventing system:

- **Event source**: Primary event producers and can also act as a link between event producers and consumers.

- **Broker**: Resources that can collect a group of cloud events. They provide a discoverable HTTP endpoint for receiving events. The producers or sources send events to brokers via HTTP POST calls to this endpoint. Knative supports different types of brokers.

- **Triggers**: Triggers are a way to filter cloud events from a broker and send it to a sink/Knative service endpoint (using HTTP POST).

- **Channel**: Event delivery mechanism that can deliver events to multiple destinations via subscriptions.

- **Subscription**: Defines the channel and the sink/subscriber the event should be delivered to.

- **Event registry**: Catalog of event types brokers can consume. This is used when triggers are used with brokers.

To reiterate the concepts visually, we have the following diagram:

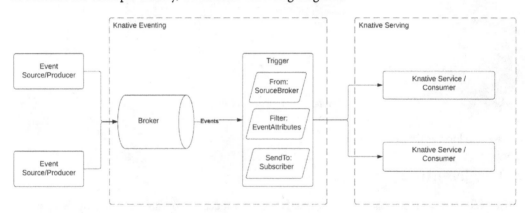

Figure 7.5 – Eventing system

That covers the Knative Eventing component. In the following section, we will briefly learn about the Kubernetes building blocks that make Knative-like frameworks possible – CRDS and operators.

Kubernetes custom resources and resource definitions

Most of the components we referred to are **custom resources (CRs)** in a Kubernetes environment and are deployed as **custom resource definitions (CRDs)**. Let's understand what CRs and CRDs are before we jump into the practical.

Kubernetes systems are collections of entities called objects and they collectively represent the state of the Kubernetes cluster. Objects, as described by Kubernetes, are *records of intent*, and Kubernetes controllers work to ensure any object that is defined will exist in the Kubernetes cluster in the desired state. Objects will fall into a resource type that fundamentally describes the properties of an object. For every resource, there is a Kubernetes API endpoint that can perform CRUD operations. This allows objects of the said resource type to be created, read, updated, and deleted. For example, a Pod is a resource type and an application Pod you create is an object of the type Pod.

A CR is an extension to the core Kubernetes API, allowing users to define their own resource type for special use cases. Since they are not part of the default cluster installations, they need to be registered dynamically with the cluster. A CRD is the means by which you declaratively add a CR to the Kubernetes API. In some use cases, creating CRDs is good enough, but in some cases, you might also want to create operators/controllers (control loops that constantly monitor the state of one or more resource types and converge the Kubernetes state to match that).

Knative and service meshes

A service mesh is an infrastructure component that can be used to facilitate communication between various services (or applications) in a Kubernetes environment as well as with the outside world. It provides traffic routing, observability, and network security in a way that is transparent to the application. Inter-service communication has always been achieved in the microservice world without a service mesh, by each developer team understanding and sticking to a common standard for how their service would expose their endpoints and receive traffic. But this becomes unmanageable and error-prone as the number of services grows. This is where the service mesh comes in. There are a number of projects that offer service meshes, most notably Istio, which was an open source project created by Google and IBM. A service mesh architecture is shown in the following figure:

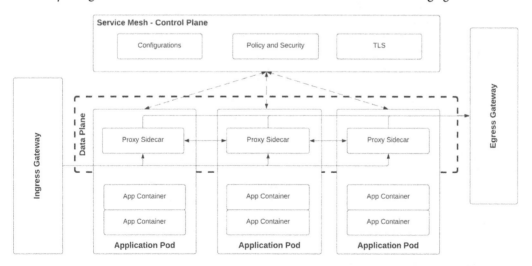

Figure 7.6 – Service mesh

While part of the service mesh's functionality is to provide communication between services, they also are used to provide load balancers/endpoint gateways for traffic that is coming into the Kubernetes cluster. This feature/service is usually called an ingress gateway. One of the advantages of using Knative is the simplified traffic routing and scaling. Knative depends on an ingress gateway or service mesh to achieve this. In earlier versions of Knative, Istio was the de-facto gateway used. But that has changed and now there are multiple supported gateways. One of them is Kourier, which is an ingress gateway project under the Knative umbrella. To learn more about the concepts of Service Mesh, please visit the Istio documentation here: `https://www.redhat.com/en/topics/microservices/what-is-a-service-mesh`.

Knative installation and setup

Now that we understand the Knative framework, let's install and set up Knative on top of our minikube cluster.

Client installation

We already have a Kubernetes cluster created using minikube; we will continue using that cluster for deploying Knative. To begin, let us install the Knative CLI, which makes a developer's life easier by simplifying Knative management:

1. Download the appropriate binary from `https://github.com/knative/client/releases/`:

```
safeer@serverless101:~/temp$ curl -sL https://github.com/
knative/client/releases/download/knative-v1.3.1/kn-linux-amd64
-o ./kn
safeer@serverless101:~/temp$ chmod a+x kn
safeer@serverless101:~/temp$ sudo mv kn /usr/local/bin/
safeer@serverless101:~/temp$ kn version
Version:  v1.3.1
....OUTPUT TRUNCATED....
```

2. Knative provides a quick start subcommand as a plugin to the Knative CLI. If you don't have minikube set up already, you can use this plugin. Install it as follows:

```
safeer@serverless101:~/temp$ curl -sL https://github.com/
knative-sandbox/kn-plugin-quickstart/releases/download/
knative-v1.3.1/kn-quickstart-linux-amd64 -o kn-quickstart
safeer@serverless101:~/temp$ chmod a+x kn-quickstart
safeer@serverless101:~/temp$ sudo mv kn-quickstart /usr/local/
bin/
safeer@serverless101:~/temp$ kn quickstart version
Version:  v1.3.1
Build Date:   2022-03-14 15:50:09
Git Revision: 3ca6b55
```

3. Initiate the cluster and Knative setup:

```
safeer@serverless101:~/temp$ kn quickstart minikube
```

4. While the quick start is an easy way to do the installation, we will use the normal way.

   Install the Serving component CRDs:

```
safeer@serverless101:~$ kubectl apply -f https://github.com/
knative/serving/releases/download/knative-v1.3.0/serving-crds.
yaml
customresourcedefinition.apiextensions.k8s.io/certificates.
networking.internal.knative.dev created
....TRUNCATED.......
customresourcedefinition.apiextensions.k8s.io/images.caching.
internal.knative.dev created
```

5. Install the Serving component core:

```
safeer@serverless101:~$ kubectl apply -f https://github.com/
knative/serving/releases/download/knative-v1.3.0/serving-core.
yaml
namespace/knative-serving created
clusterrole.rbac.authorization.k8s.io/knative-serving-
aggregated-addressable-resolver created
clusterrole.rbac.authorization.k8s.io/knative-serving-
addressable-resolver created
...TRUNCATED - MUCH LARGER OUTPUT THAN CRDs.....
validatingwebhookconfiguration.admissionregistration.k8s.io/
validation.webhook.domainmapping.serving.knative.dev created
validatingwebhookconfiguration.admissionregistration.k8s.io/
validation.webhook.serving.knative.dev created
secret/webhook-certs created
```

6. Once the components are successfully installed, you need to install a networking layer for serving. This is usually a service mesh or ingress gateway. While Istio was the default previously, Kourier has become a Knative project and we will be installing that. First, we will install the Kourier controller:

```
safeer@serverless101:~$ kubectl apply -f https://github.com/
knative/net-kourier/releases/download/knative-v1.3.0/kourier.
yaml
namespace/kourier-system created
configmap/kourier-bootstrap created
....TRUNCATED......
service/kourier created
service/kourier-internal created
```

7. All Knative components are installed in a separate namespace called `knative-serving`. We will update the network configs to have Kourier as the ingress controller:

```
safeer@serverless101:~$ kubectl patch configmap/config-network \
>    --namespace knative-serving \
>    --type merge \
>    --patch '{"data":{"ingress.class":"kourier.ingress.
networking.knative.dev"}}'
configmap/config-network patched
```

8. Check whether all the Knative Kerving components are running:

```
safeer@serverless101:~$ kubectl get pods -n knative-serving
NAME                          READY   STATUS   RESTARTS   AGE
activator-855fbdfd77-
qs5zb              1/1    Running   0          20m
autoscaler-85748d9cf4-
dr6mt              1/1    Running   0          20m
controller-798994c5bd-
w8gkd              1/1    Running   0          20m
domain-mapping-59fdc67c94-
zjgl5        1/1    Running   0         20m
domainmapping-webhook-6df595d448-
q9kfn    1/1    Running   0         20m
net-kourier-controller-74dc74797-
xgfck    1/1    Running   0         8m38s
webhook-69fdbbf67d-
ttjlg              1/1    Running   0          20m
```

9. We are going to configure the default domain for Knative to use `sslip.io`. Now, `sslip.io` is a system where a query to the `<IPv4Address>.sslip.io` subdomain will return the same IP. This suffix system helps Knative to provide resolvable DNS endpoints:

```
safeer@serverless101:~$ kubectl apply -f https://github.com/
knative/serving/releases/download/knative-v1.3.0/serving-
default-domain.yaml
job.batch/default-domain created
service/default-domain-service created
```

10. Next, we will deploy a Knative Serving sample application. While we can do this with kubectl and YAML files as before, let's try out the Knative CLI and see how easy it is to use it. This sample application reads a name to greet from the TARGET environment variable and returns a `Hello <TARGET>!` message:

```
safeer@serverless101:~/temp$ kn --cluster minikube service
create hello --image gcr.io/knative-samples/helloworld-go --port
8080 --env TARGET=Knative
Creating service 'hello' in namespace 'default':
```

```
<<TRUNCATED OUTPUT>>
 51.082s Ready to serve.

Service 'hello' created to latest revision 'hello-00001' is
available at URL:
http://hello.default.10.109.65.82.sslip.io
```

11. Access the application:

```
safeer@serverless101:~/temp$ curl http://hello.
default.10.109.65.82.sslip.io
Hello Knative!
```

12. List the Knative services:

```
safeer@serverless101:~/temp$ kn --cluster minikube service list
NAME     URL     LATEST     AGE     CONDITIONS     READY     REASON
hello    http://hello.default.10.109.65.82.sslip.io     hello-
00001    2m6s    3 OK / 3     True
```

13. As you can see in the previous output, we were able to deploy a "hello world" application and retrieve its URL with minimal configuration. Let us inspect the Knative service we just deployed a bit more closely:

```
safeer@serverless101:~/temp$ kn --cluster minikube service
describe hello
Name:       hello
Namespace:  default
Age:        48m
URL:        http://hello.default.10.109.65.82.sslip.io

Revisions:
   100%  @latest (hello-00001) [1] (48m)
      Image:    gcr.io/knative-samples/helloworld-go (pinned to
5ea96b)
      Replicas:  0/0

Conditions:
   OK TYPE                 AGE REASON
   ++ Ready                47m
   ++ ConfigurationsReady  47m
   ++ RoutesReady          47m
```

14. Note that a Knative service is not the same as a Kubernetes service; it is a different resource type called `KService` or `ksvc`. You can use kubectl to get details of it:

```
safeer@serverless101:~/temp$ kubectl get ksvc
NAME     URL      LATESTCREATED    LATESTREADY    READY    REASON
hello    http://hello.default.10.109.65.82.sslip.io    hello-
00001        hello-00001    True
```

As you can see, the output of the Knative command that described the hello `ksvc` shows some basic application details. There are two important details under `Revisions`: the traffic distribution split and the revision to which the traffic is going. Since this is a default installation, 100% of the traffic goes to the latest revision.

In the following program, I am going to show some commands to manage the applications. The outputs will be trimmed down or completely avoided to save space and better readability.

Update the service by changing the value of the `TARGET` environment variable:

```
safeer@serverless101:~/temp$ kn --cluster minikube service update
hello --env TARGET="Knative v2"

safeer@serverless101:~/temp$ curl http://hello.default.10.109.65.82.
sslip.io
Hello Knative v2!
```

Notice the revision:

```
safeer@serverless101:~/temp$ kn --cluster minikube service describe
hello|grep -A 1 -E '^Revisions'
Revisions:
  100%  @latest (hello-00002) [2] (5m)

safeer@serverless101:~/temp$ kn --cluster minikube revision list
NAME      SERVICE    TRAFFIC    TAGS    GENERATION    AGE    CONDITIONS
    READY    REASON
hello-00002    hello    100%           2          6m37s    3 OK / 4 True
hello-00001    hello                   1          64m 3 OK / 4 True
```

Here are some other useful commands:

- List the revisions:

  ```
  kn --cluster minikube revision list
  ```

- Get more details on a specific revision:

  - `kn revision describe hello-00002`

- List the routes:

```
kn --cluster minikube route list
```

- Retrieve the service URL:

```
kn service describe greeter -o url
```

You can split the traffic between multiple revisions as follows:

```
kn --cluster minikube service update hello --tag=hello-00002=green
kn --cluster minikube service update hello --tag=hello-00001=blue
kn --cluster minikube service update hello --traffic blue=80,green=20
```

There are a lot more advanced options that we can use with Serving and Eventing, but unfortunately, we can't cover all of them here. You can refer to the official documentation for more details:

- `https://knative.dev/docs/getting-started/` is the best place to start

- Red Hat offers Knative as a service and has good practical documentation at `https://redhat-developer-demos.github.io/knative-tutorial/knative-tutorial/index.html`

- `https://knative.tips/` is another good resource to learn more about the internals

While Knative is a great platform to deploy a serverless workload, you will have noticed that it doesn't have a construct to offer **Function-as-a Service** (**FaaS**) features. This is intended and the goal of Knative is to facilitate the deployment and management of stateless services. The next technology that we are going to explore is built exactly for FaaS use cases. Let us try and learn more about it in the next section.

OpenFaaS

OpenFaaS (`https://www.openfaas.com/`) is an open source project that provides a framework for building and deploying FaaS. Their tagline is *Serverless Functions Made Simple* and the claim is justifiable. The functions are built on Docker containers and orchestration can be either Kubernetes or VMs (with some caveats). It provides standard FaaS platform features such as autoscaling, auto-healing, and resource control.

OpenFaaS aims to make life easy for serverless and FaaS developers; some of the features that help in achieving this are as follows:

- Portability – create functions without vendor lock-in

- Support all languages and runtimes as well as binaries and scripts

- Built-in UI portal, versatile CLI, and simplified installation and management

- Scalable – can scale down to zero

- Faster cold start time

- Active user community

- Community and Pro editions

- Function stores

- Templates

Let us look at the OpenFaaS architecture next.

OpenFaaS architecture

OpenFaaS is written in Golang. It provides UI/API and CLI to interact with the core services. Let us look at how the architecture is set up:

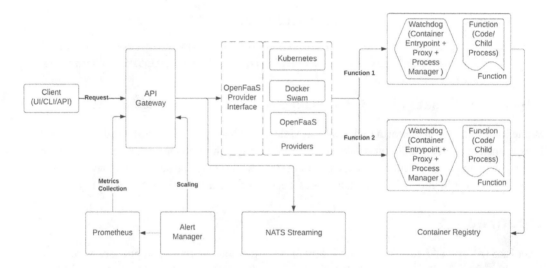

Figure 7.7 – OpenFaaS architecture

Let us look at the individual components:

- **OpenFaaS gateway**: The entry point to OpenFaaS is the gateway. It hosts all the control plane APIs as well as the API gateway for all the functions deployed. In addition, it also hosts a UI portal for OpenFaaS operations. The gateway takes care of scaling the function up or down (to and from zero).

- **FaaS provider**: A FaaS provider is the backend service for a container orchestration platform with a unified interface exposed to the gateway. This helps in supporting multiple container orchestration providers while keeping the business logic and interaction between the gateway and provider constant. There is a Golang SDK provider by OpenFaaS that can be used to write your own providers. It also provides CRUD operations for functions and the capability to invoke the function.

- **Prometheus**: Prometheus is an open source systems monitoring and alerting toolkit originally built at SoundCloud. It collects and stores its metrics as time series data and optional metadata as key-value pairs called labels. The gateway uses Prometheus to track traffic and other infrastructure metrics.

- **Alert Manager**: Alert Manager is a project created by Prometheus to support alerting. It receives alerts from systems such as Prometheus, then performs operations such as grouping and de-duplicating and then routing them to the appropriate destination. The gateway uses this in conjunction with Prometheus to get notified when traffic to a function breaches the upper or lower limits set and uses that information to scale up or down accordingly.

- **Kuberenetes/container orchestrators**: While Kubernetes is the preferred orchestration backend for OpenFaaS, it also supports Docker Swarm, Hashicorp Nomad, and cloud-vendor-provided Kubernetes services such as GKE, EKS, and Fargate.

- **NATS streaming**: NATS is a secure and simple distributed messaging system (`https://nats.io/`). It fits most cloud messaging and IoT/edge communication use cases. It is used to asynchronously execute functions.

- **Function container**: Containerized function code and a binary called a watchdog as the container entry point.

- **Function watchdog**: A sidecar/entry point for controlling and monitoring the function.

- **Container registry**: A registry is required to push the images built as well as deploy them at scale using OpenFaaS.

Now let us look at the function runtime and why watchdog is important.

Function runtime and watchdog

The functions are run as Docker/OCI-compliant containers. The additional requirements from OpenFaaS to run these containers are limited. The container should have an HTTP server listening for communication, assume no persistent storage, and of course be stateless, unless the state can be managed in an external data system. The HTTP server becomes the container entry point and is responsible for running the function code along with input-output processing.

While the need for an HTTP server in the container is a fundamental requirement, OpenFaaS makes it easy for the developer by providing a ready-made solution called a watchdog. A watchdog is a lightweight HTTP server written in Golang. The watchdog will have an understanding of how to execute the function's business logic. Hence a combination of the code and the watchdog completes the container runtime image. There are two flavors of watchdog, *classic* and the newer generation, *of-watchdog*.

Classic watchdog

The container is launched when a function is deployed or scaled. As the entry point process for the container, the watchdog starts and waits for incoming HTTP requests via the OpenFaaS gateway. The classic watchdog has the following lifecycle:

1. Read the incoming HTTP request.
2. Fork the actual function, called a handler (this usually is a binary or a script and its programming runtime).
3. Pass the request to the STDIN of the function.
4. Wait for the function to finish processing the request.
5. Relay the STDOUT and STDERR of the function execution back as an HTTP response.
6. Once the response is sent out, the function process exits, but the watchdog stays up.
7. When a new request is received, the same process is repeated.

Next, let us look at the of-watchdog.

Of-watchdog

Like the classic watchdog, this one also spins up its HTTP server when the container is deployed. It will also spawn the function code the same way. But unlike the classic watchdog, the of-watchdog expects the process to accept HTTP requests and responses and not STDIN/OUT prompts. It also expects the function to be a long-running process rather than one that dies after one request. The of-watchdog acts like a reverse proxy passing requests between the FaaS gateway and the function web server.

The of-watchdog is not a replacement for the classic watchdog. In some cases where either the programming language or the nature of the business use case does not support/need an HTTP server, classic is the only option. Classic is also extremely simple, but spinning up a new process every time a web request is received is expensive. This also allows the function to persist state across multiple request invocations. Another notable advantage is that unlike STDOUT, it can stream output, and this helps if the output is larger than the memory allocated to the function:

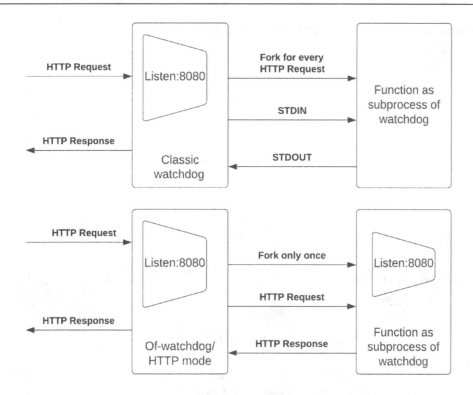

Figure 7.8 – Watchdogs

Next, let us look at the asynchronous variants of OpenFaaS functions.

Asynchronous functions

Normal FaaS functions are synchronous; they respond to requests and return the results immediately. But OpenFaaS also offers asynchronous functions. This allows OpenFaaS to run long-running or background tasks using NATS streaming. The workflow is as follows:

1. A request is sent to the asynchronous HTTP endpoint of the function on the gateway.

2. The gateway serializes the HTTP request into the NATS streaming format and sends it to NATS.

3. A queue worker picks up the request from the NATS queue, deserializes it, and uses it to invoke the function.

4. Optionally, a callback URL can be registered while placing the asynchronous request. This will allow the function to post back the response through the callback URL.

The following diagram shows the call path of asynchronous workflow:

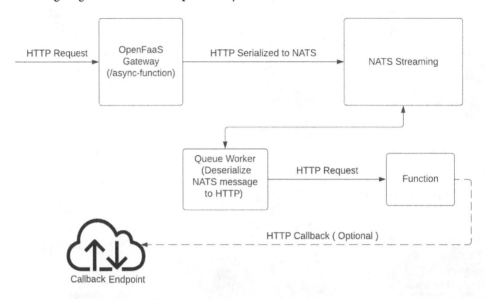

Figure 7.9 – Asynchronous function calls

We have covered most of the infrastructure and architecture components of OpenFaaS. Let us quickly look at a simplified and powerful version of standalone OpenFaaS next.

FaaSD

FaaSD is a lightweight and portable implementation of the core OpenFaaS functionalities. It is great for software development environments as well as production deployments where resources are scarce. IoT and edge cases are good use cases for FaaSD. It runs on a single host and uses the containerd runtime and **container network interface (CNI)** for orchestration infrastructure. This removes a lot of the maintenance burden from developers and operators. FaaSD is also extremely fast.

OpenFaaS installation and setup

There are multiple ways to install OpenFaaS. The recommended one is to use the official CLI: `arkade`. We will take this route.

1. We'll start by installing `arkade`:

    ```
    root@serverless101:~# curl -SLsf https://get.arkade.dev/ | sudo
    sh
    Downloading package
    …..TRUNCATED…..
    Git Commit: fbb376b431a2614be063d8a3b21b73ad01d53324
    ```

2. Let's now install OpenFaaS with `arkade` on our existing minikube cluster. Please note that you will have to install the actual kubectl utility as `arkade` doesn't seem to pick up our minikube alias for kubectl:

```
safeer@serverless101:~$ arkade install openfaas
Using Kubeconfig: /home/safeer/.kube/config
Client: x86_64, Linux
….TRUNCATED…..
================================================================
========
= OpenFaaS has been installed.                                 =
….OUTPUT TRUNCATED..
Thanks for using arkade!
```

3. This command has a very long output and is truncated. The output mostly contains the instructions to set up and use OpenFaaS. We will cover those commands as well, shown here. Now verify that the installation is successful:

```
safeer@serverless101:~$   kubectl -n openfaas get deployments -l
"release=openfaas, app=openfaas"
NAME              READY    UP-TO-DATE    AVAILABLE    AGE
alertmanager       1/1    1                1            58s
basic-auth-plugin  1/1    1                  1          58s
gateway            1/1    1              1              58s
nats               1/1    1                1            58s
prometheus         1/1    1                1            58s
queue-worker       1/1    1                1            58s
```

4. Let's now install the OpenFaaS CLI:

```
safeer@serverless101:~$ curl -SLsf https://cli.openfaas.com |
sudo sh
Finding latest version from GitHub
0.14.2
Downloading package https://github.com/openfaas/faas-cli/
releases/download/0.14.2/faas-cli as /tmp/faas-cli
Download complete.
…TRUNCATED….
 version: 0.14.2
```

5. Now, let us set up the OpenFaaS API gateway and expose it:

```
safeer@serverless101:~$ kubectl rollout status -n openfaas
deploy/gateway
deployment "gateway" successfully rolled out
```

6. From another window, run the following:

```
safeer@serverless101:~$ kubectl port-forward -n openfaas
--address 0.0.0.0 svc/gateway 8080:8080
Forwarding from 0.0.0.0:8080 -> 8080
Handling connection for 8080
```

7. Return to the previous shell and get the password to do basic auth with the CLI or UI. The default username is `admin`:

```
safeer@serverless101:~$ PASSWORD=$(kubectl get secret -n
openfaas basic-auth -o jsonpath="{.data.basic-auth-password}" |
base64 --decode; echo)
safeer@serverless101:~$ echo $PASSWORD
1BL3E2OYf67CWr6i4RXxBB8S9
```

8. Log in to the CLI using the following command:

```
safeer@serverless101:~$ echo -n $PASSWORD | faas-cli login
--username admin --password-stdin
Calling the OpenFaaS server to validate the credentials...
credentials saved for admin http://127.0.0.1:8080
```

9. Now, we are ready to create and deploy functions. OpenFaaS makes it easy to create functions by providing a boilerplate function and container runtime definition as a template. The OpenFaaS CLI has a template engine that can convert this template into a directory structure with an empty function handler and the associated watchdog and container runtime information. The public and community curated templates are available in the following GitHub repo: https://github.com/openfaas/templates. This is the default template store that `faas-cli` talks to. Developers can also create their own template stores and override the default one in the CLI. Let us start by listing templates available in the default store:

```
safeer@serverless101:~/openfaas-workspace$ faas-cli template
store ls -v

go                      Go      x86_64    openfaas      Classic
Golang template
java8                   Java    x86_64    openfaas      Java 8
template
java11                  Java    x86_64    openfaas      Java 11
template
java11-vert-x           Java    x86_64    openfaas      Java 11
Vert.x template
...TRUNCATED....
```

10. Let us create a working directory for OpenFaaS development and pull the templates from the default store:

```
safeer@serverless101:~$ mkdir openfaas-workspace/;cd openfaas-
workspace/
```

11. Pull the default templates to the local working directory:

```
safeer@serverless101:~/openfaas-workspace$ faas-cli template
pullhttps://github.com/openfaas/templates.git
```

12. We will use the python3 Debian template; let us first check its directory structure:

```
safeer@serverless101:~/openfaas-workspace$ tree template/
python3-debian/
template/python3-debian/
├── Dockerfile
├── function
│   ├── __init__.py
│   ├── handler.py
│   └── requirements.txt
├── index.py
├── requirements.txt
└── template.yml
```

13. Now that we have templates available, let us create our first function in python3. We will name the function firstpy. The following command creates a new function from the template:

```
safeer@serverless101:~/openfaas-workspace$ faas-cli new firstpy
--lang python3-debian
Folder: firstpy created.
Function created in folder: firstpy
Stack file written: firstpy.yml
```

14. Let us examine the function definition YAML as well as the directory structure:

```
safeer@serverless101:~/openfaas-workspace$ tree firstpy
firstpy
├── __init__.py
├── handler.py
└── requirements.txt
0 directories, 3 files
safeer@serverless101:~/openfaas-workspace$ cat firstpy.yml
version: 1.0
provider:
  name: openfaas
  gateway: http://127.0.0.1:8080
```

```
functions:
  firstpy:
    lang: python3-debian
    handler: ./firstpy
    image: firstpy:latest
```

15. Here is the function handler template:

```
safeer@serverless101:~/openfaas-workspace$ cat firstpy/handler.
py
def handle(req):
    """handle a request to the function
    Args:
    req (str): request body
    """

    return req
```

16. Now let us add our code and build the function. Before we build, edit the firstpy.yml file and modify your image name, and prepend your Docker username to it. Make sure you are logged in to your Docker account using the docker login command as well:

```
safeer@serverless101:~/openfaas-workspace$ cat firstpy/handler.
py
import json
def handle(req):
    return json.dumps({"message" : req })

safeer@serverless101:~/openfaas-workspace$ faas-cli build -f ./
firstpy.yml
[0] > Building firstpy.
...OUTPUT TRUNCATED...
[0] < Building firstpy done in 38.31s.
[0] Worker done.

Total build time: 38.31s

safeer@serverless101:~/openfaas-workspace$ docker images|grep
firstpy
firstpy
latest    dd385dcccf0b    About a minute ago    936MB

safeer@serverless101:~/openfaas-workspace$ faas-cli push -f ./
firstpy.yml
```

```
[0] > Pushing firstpy [safeercm/firstpy:latest].
The push refers to repository [docker.io/safeercm/firstpy]
...OUTPUT TRUNCATED....
[0] < Pushing firstpy [safeercm/firstpy:latest] done.
[0] Worker done.
```

17. As you can see, the container is successfully built and pushed to Docker Hub. There are ways to use a private/local registry instead, but covering that is outside the scope of this tutorial:

```
safeer@serverless101:~/openfaas-workspace$ faas-cli deploy -f ./
firstpy.yml
Deploying: firstpy.

Deployed. 202 Accepted.
URL: http://127.0.0.1:8080/function/firstpy

safeer@serverless101:~/openfaas-workspace$ faas-cli list
Function                 Invocations    Replicas
firstpy                      0               1

safeer@serverless101:~/openfaas-workspace$ faas-cli describe
firstpy
Name:                  firstpy
Status:                Ready
Replicas:              1
Available replicas:    1
Invocations:           3
Image:                 safeercm/firstpy:latest
Function process:      python3 index.py
URL:                   http://127.0.0.1:8080/function/firstpy
Async URL:             http://127.0.0.1:8080/async-function/
firstpy
Labels                 faas_function : firstpy
                       uid : 809905608
Annotations            prometheus.io.scrape : false
```

18. As you can see, the function is deployed and available via the OpenFaaS gateway. Let us send a message to it and see whether it returns the JSON:

```
safeer@serverless101:~/openfaas-workspace$ curl
http://127.0.0.1:8080/function/firstpy -d "OpenFaaS is awesome"
{"message": "OpenFaaS is awesome"}
```

19. As you can see, the function is returning the response. We can also invoke the function from the CLI as follows:

```
safeer@serverless101:~/openfaas-workspace$ faas-cli invoke
firstpy
Reading from STDIN - hit (Control + D) to stop.
Serverless is fun :)
{"message": "Serverless is fun :)\n"}
```

This tutorial should have given you a basic idea of how to use OpenFaaS. For more information, check out the official OpenFaaS training at https://www.edx.org/course/introduction-to-serverless-on-kubernetes.

Example project – implementing a GitHub webhook with a Telegram notification

As the title suggests, this project is about using GitHub webhooks and notifications using Telegram. A webhook is a passive way for an application to communicate with another one. The app that is meant to receive the notification will expose a web endpoint, and the app initiating the communication will post data to that endpoint in a format specified by the receiver, optionally with authentication and/or authorization. Once the notification is received, the receiver can process it according to its business logic.

In this case, what we are doing is making use of GitHub's webhooks facility. This allows us to receive a notification payload from GitHub whenever a specific event occurs to a repo or organization that we own. In our case, we want to get notified whenever a push happens to one of our repos. GitHub will send a JSON payload with a lot of information about the push, including repo details, push and commit details, and so on. What we intend to do after receiving the payload is to extract some key information from the payload and send it as a Telegram message to the repo owner.

High-level solution

The objective is to notify a user via Telegram whenever they push some commits to a GitHub repo. The entire workflow is as follows:

1. The user pushes some changes to a GitHub repo.
2. The push will trigger a webhook call.
3. GitHub will post the corresponding payload to the OpenShift function endpoint.
4. The function will process the payload and post it to the Telegram bot API.
5. The bot will send an instant message to the user about the push.

This entire workflow usually finishes in seconds. In the *Design and architecture* section that follows, we will inspect the resources we need as well as the architecture of the solution.

Design and architecture

Implementing this will require multiple resources/services:

- **GitHub**

  A repo – public or private – that you own.

  Configure a webhook to send a payload whenever a push to this repo takes place.

- **Telegram**

  Create a bot with your account.

  Obtain an auth token and your account's chat ID.

- **Serverless**

  A Knative Serving application that exposes an endpoint. This will be the webhook endpoint that will be configured in GitHub.

When a payload is received, the app parses the JSON payload and uses the Telegram bot API to send a chat notification to you.

We will choose Red Hat OpenShift as our serverless platform. OpenShift is a managed Kubernetes offering and is used by large enterprises. OpenShift has adopted Knative as its serverless offering on top of OpenShift. Red Hat calls their managed serverless offering OpenShift Cloud Functions. This is what we will be using to run our webhook, and it will be written in Python.

You can see the architecture and workflow of the webhook we are designing here:

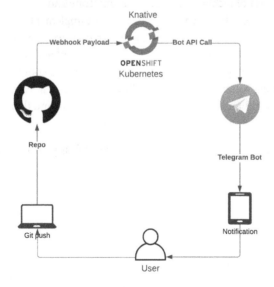

Figure 7.10 – GitHub webhook architecture

The workflow is very straightforward, as you have already seen. Let us see how to implement it next.

Application code and infrastructure automation

The application/OpenShift function will be written in Python. To start running the application, we need to sign up and set up our OpenShift account. Red Hat provides a one-month sandbox free of charge. We will use this Kubernetes cluster and the serverless offering that comes with it. The code and instructions will be available in this book's GitHub repo at the following location:

```
https://github.com/PacktPublishing/Architecting-Cloud-Native-
Serverless-Solutions/tree/main/chapter-7
```

Summary

Kubernetes has revolutionized the way infrastructure and applications are deployed and managed. It has abstracted out a lot of complexity and manual processes involved in such operations. But along with the ease came its own complex ecosystem and processes. Where serverless and FaaS frameworks shine is when they utilize the power of Kubernetes and shield the developer and operators from additional manual processes and complexity.

In this chapter, we learned the fundamentals and gained an overview of Docker and Kubernetes to get some foundational knowledge that was essential to understand the frameworks we covered in this chapter. We then started understanding Knative, the open source serverless framework that powers the CloudRun serverless platform of Google Cloud. We got an understanding of the building blocks that should be part of a serverless framework. But it is also important to note that Knative is not a FaaS platform; rather, it is a set of technologies used to orchestrate and deploy stateless microservices without compromising developer velocity. FaaS, of course, can be implemented easily on Knative, but the function is not a primary construct in its ecosystem. After Knative, we moved on to OpenFaaS, which was built to run functions on any platform with ease of development and deployment. Infrastructure components are not hard to deploy or understand with OpenFaaS. To complete the learning, we also created a sample project using Knative on OpenShift. This completes our learning for this chapter, but our coverage of the frameworks and technologies was not exhaustive and there are more frameworks and stacks that can improve the serverless story of Kubernetes.

In the next chapter, we will cover another cloud-agnostic serverless platform – Apache OpenWhisk.

8

Self-Hosted FaaS with Apache OpenWhisk

Serverless has revolutionized cloud computing, and we have talked about a number of vendors and technology platforms in the previous chapters. The central theme of all serverless platforms has been **Function as a Service** (**FaaS**) as the coordinating and controlling technology. Though AWS pioneered FaaS and other cloud vendors caught on, it shouldn't be limited to a few cloud vendor offerings. Even a single technology platform such as Kubernetes should not be the only way to deploy serverless or FaaS, even though it is open source and vendor-neutral. There have been multiple initiatives to create a vendor-neutral and platform-agnostic serverless platform, some more successful than others. Apache OpenWhisk has evolved as one of the top platforms of this kind.

OpenWhisk started off as a serverless project in IBM in 2015, codenamed **whisk**, shortly after AWS Lambda was announced. While the fundamental idea was to power an IBM Cloud FaaS offering using this project, they soon decided to open source and grant the power of serverless functions to the community. The project was renamed OpenWhisk to indicate its open source status. By the end of 2016, IBM had donated the codebase to Apache Foundation, and it became an incubating project. By the end of 2019, OpenWhisk graduated to the status of an Apache Project from its incubator designation.

We will cover the following in this chapter:

- OpenWhisk – concepts and features
- Architecture
- Creating and managing actions and entities
- Administration and deployments
- Project – implementing stream processing

OpenWhisk – concepts and features

Apache OpenWhisk calls itself "*an open-source, distributed serverless platform that executes functions (fx) in response to events at any scale.*" OpenWhisk manages the platform, its scaling, and other operational aspects, freeing the developer to do what they do best: deliver business-critical applications. In this case, they would be serverless applications, though. The fundamental programming model of OpenWhisk is that the developers can write business logic as a function code in any of the programming languages supported. These functions are then invoked in response to events or schedules.

Let us look at the core concepts and building blocks of this programming model:

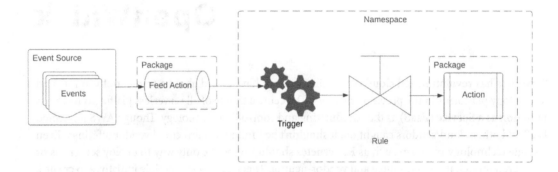

Figure 8.1 – OpenWhisk programming model

Now let us look into these entities in detail.

Actions and action chains

OpenWhisk is a FaaS platform, and within it, the function is referred to as an **action**. This is the computation component of OpenWhisk that will respond to and process *events*. Actions should be stateless, and any persistence that is required should be achieved with an external data store. Being stateless makes actions highly scalable. Actions are triggered by events – we will talk about events in the next section. Events can also be manually triggered using the OpenWhisk API/CLI.

If a business workflow requires multiple phases of processing, each depending on the outcome of the previous phase, actions can be chained. When chained, the output of each action – in JSON format – is passed on as input to the next action. This chaining is called a **sequence** and OpenWhisk allows you to define these sequences as an entity and then use the sequences like unified/single actions and invoke them as such. Practically, you will define a sequence as an action, with an ordered list of actions as its execution logic rather than an action file. You can also build more complicated workflows using rules, which we will see in the coming sections. The following diagram shows the structure of a sequence:

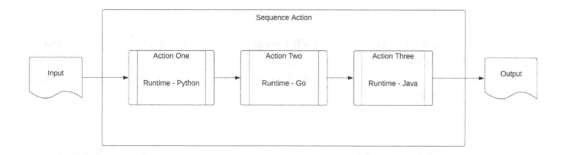

Figure 8.2 – Action sequence

As you can see in the diagram, the sequence is very useful to chain multiple actions and process data accordingly. Next, let us look at a few other entities in the OpenWhisk system.

Events, sources, and feeds

The entire OpenWhisk concept is built around events. As defined previously, an event is the occurrence of anything important to our business or software infrastructure. Events can be from any source and of any type. Some common examples of event sources are as follows:

- Message queues
- **Change capture data (CDC)** systems
- Databases
- Web applications and inter-application communication
- API calls
- Internet of Things and edge devices
- Timers
- Feeds and notifications for SaaS applications

Next, let us look at triggers, which are useful for binding events and functions.

Triggers

A trigger is basically an intention that specifies that you want to respond to a class of events. This could be any category of events, such as timers, geolocation updates, file upload to cloud storage, and so on. These intentions/triggers are named so that they can be associated with feeds and rules.

Feeds and feed actions

Feeds are pieces of code that are associated with event sources and help in triggering action/computation code. An action is only invoked when an event occurs. The code that achieves this is called a feed action. It is responsible for creating, deleting, pausing, and resuming the event stream that is associated with a feed.

Rules

Rules allow triggers to be associated with one or more actions. Once an association is made, each time a matching event coming from the event source, a trigger is activated, which in turn invokes the actions associated with them. The following diagram describes how rules bind events and action via triggers:

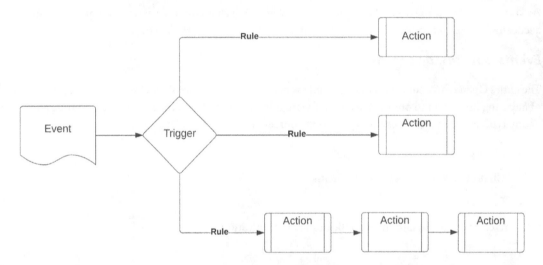

Figure 8.3 – One trigger and multiple actions

It is possible for one trigger to be associated with multiple rules, which will invoke multiple functions when the trigger is fired. Similarly, multiple triggers can be associated with multiple rules, but all of the rules point to one action. The following diagram describes multiple events associated to the same action:

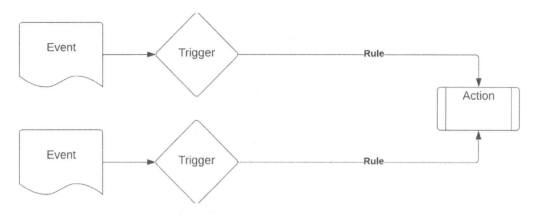

Figure 8.4 – Multiple triggers and one action

Next, we will look at a few higher-level constructs, such as namespaces and packages, in OpenWhisk.

Namespaces

Namespaces are logical containers that can include other OpenWhisk resources such as actions, feeds, packages, and so on. It is an administrative construct and is the top-level entity in the OpenWhisk programming model. Namespaces and other Openwhisk entities are named like Unix file paths or URL paths, with a leading forward slash followed by the name of the namespace where the entity belongs. The `/whisk.system` namespace is reserved for entities that are distributed with the OpenWhisk system.

Packages

Packages are a way to bundle your actions and share them with others. Ideally, a set of related actions will be packaged together. A package can include actions as well as feeds (which are special-purpose actions). A package can be used to organize the code and can be versioned like any software package. A package can be marked as shareable and users in another namespace can bind that package to their namespace, given that they know the full name of the package (under the correct namespace).

Architecture

OpenWhisk uses a number of open source technologies to power its platform. These include Nginx, CouchDB, Docker, and Kafka. The architecture diagram is as follows:

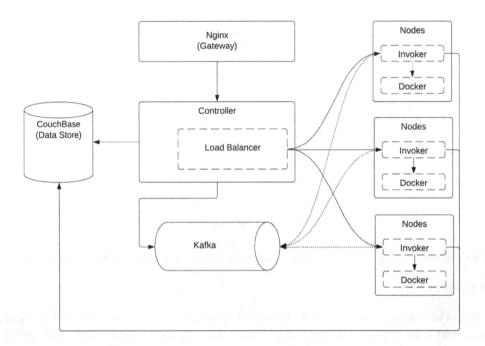

Figure 8.5 – OpenWhisk architecture

Let us examine the components one by one:

- **Nginx**: Nginx is a popular and high-performance HTTP reverse proxy. It acts as the entry point to the OpenWhisk system and proxies all of its requests to the backend. It can also be used for offloading SSL termination. Since Nginx is a stateless application, it can be easily scaled according to the needs of the serverless platform.

- **Controller**: The controller is the brain of the OpenWhisk system. It receives all incoming requests to the whisk system that Nginx proxies to it and then processes them accordingly. This component is written in Scala and implements the full REST API of OpenWhisk. It is responsible for authenticating and authorizing the requests, as well as orchestration of one or more actions as required by incoming requests. The controller handles all CRUD – *create*/*read*/*update*/*delete* – requests to all OpenWhisk entities.

- **CouchDB**: Apache CouchDB is an open source JSON document store that is highly scalable. It is the persistent store that keeps the state of all things that happens in the OpenWhisk system. It stores a number of things, including user credentials, namespaces, entities, action/trigger/ rules definitions, and more.

- **Kafka**: Apache Kafka is a popular open source streaming platform that is highly scalable and distributed. It is used widely for implementing a number of streaming and Pub/Sub use cases, OpenWhisk primarily uses Kafka as a message bus for coordinating between the controller and invokers.

- **Invokers**: The invoker is an OpenWhisk service that is responsible for executing functions. Based on the resource constraints and function parameters, invokers spin up Docker containers containing function code and execute them. There will be a defined number of invokers in an OpenWhisk cluster.

- **Docker**: Actions are the unit of execution in OpenWhisk and Docker is the technology of choice for wrapping up the action code and executing it with the chosen runtime and environment. Unlike container orchestrators such as Kubernetes, in the OpenWhisk ecosystem, the controller and invoker carry out the orchestration and execution with the help of Kafka, CouchDB, and Nginx. Most of the core components of OpenWhisk itself are built and deployed as Docker containers.

Lifecycle of a request

To understand this infrastructure and how it is tied together, we need to walk through the lifecycle of a function invocation. Invoking an action is the start of everything in OpenWhisk. There are a number of ways to invoke an action, including the following:

- An action is published to the web/exposes an API and a request hits that API

- Another action (probably from chaining) invokes the action via the OpenWhisk API

- Direct invocation using the OpenWhisk CLI

- A trigger is fired and a rule associated with it invokes the action

A function can be invoked synchronously or asynchronously. Assuming a function is invoked using one of the methods listed previously, let us explore what happens within the OpenWhisk ecosystem:

1. Nginx receives the invocation request at its public endpoint. Since OpenWhisk API is RESTful, the invocation will be an HTTP POST request to `/api/v1/namespaces/<USER_NAMESPACE>/actions/<ACTION_NAME>`.

2. Nginx completes SSL termination (if configured), performs basic HTTP validation checks, and forwards the request to the controller API endpoint.

3. The controller receives the request and does a preprocessing by examining the request and deciding what to do with it. Given that this is a POST request to an action endpoint, the controller proceeds to authentication and authorization.

4. In order to authenticate the user invoking the request and to verify whether the said user has the privilege to invoke the action, the controller contacts the CouchDB and looks up a database called `subjects`.

5.   If both authentication and authorization checks pass, the next phase is to retrieve action information from the `whisks` database in CouchDB. This includes code, default parameters, and resource restrictions. The controller merges the default parameters with the parameters received from the function invocation to create a complete list of parameters to be passed to the action code.

6.   Once the code and parameters are ready, it is time to summon an invoker to carry out the action. For this, a subsystem of the controller called the load balancer is used.

7.   The load balancer keeps track of all invokers and knows which one of the invokers is free. It will pick one of the free invokers for this action.

8.   Once the choice is made, the controller crafts a message with details of the action and publishes it into Kafka. Once Kafka acknowledges the message is received, the next step depends on whether the invocation was synchronous or asynchronous.

9.   If the actions were invoked asynchronously, the controller creates an activation ID and returns it to the user with an HTTP 202 success code. This tracking ID can be used later to find out the status of the action invocation.

10. If the invocation was synchronous, the invocation HTTP request will block until the action result or an error is returned.

11. Coming back to the Kafka message that was left by the controller, the designated invoker will pick up the request and get ready to run the action.

12. The invoker users Docker to execute actions in an isolated and safe runtime environment. In order to achieve this, the invoker spawns a new Docker container with the runtime environment for the action injects the action code and executes the actions with the associated parameters. Once the action is complete, the invoker saves the result and the logs and then destroys the container.

13. The result of the action invocation is stored in CouchDB. Both the result and the logs are stored as an activation record associated with the activation ID mentioned earlier. The record is stored in the CouchDB database named activations.

14. Once the record is stored, it can be retrieved by using the activation ID that was returned earlier to the user.

This roughly outlines what happens behind the scenes when an OpenWhisk action is running. Now that we have a good understanding of the internals, let us look at how to deploy OpenWhisk.

Creating and managing actions and entities

There are multiple ways to deploy OpenWhisk on your infrastructure. Kubernetes has become a preferred choice for deployment. OpenWhisk provides Helm charts to help you deploy OpenWhisk infrastructure on Kubernetes. There are certain requirements to be satisfied for deploying on Kubernetes, though:

- The default installation needs at least one worker node with a minimum of two vCPUs and 4 GB of memory

- The Kubernetes cluster should be version 1.19 or above

- The privilege to create ingress so that the OpenWhisk gateway can be exposed

- The Kubernetes **hairpin-mode** shouldn't be **none** as OpenWhisk endpoints should be able to loop back to themselves

For more details, refer to the official documentation at `https://github.com/apache/OpenWhisk-deploy-kube`

The other deployment options are as follows:

- **QuickStart**: OpenWhisk provides a standalone version of itself. This standalone version is a Java process that exposes all the features of OpenWhisk in a single service. This is ideal for development environments.

- **docker-compose**: A slightly more advanced option, which is still good for developer environments is to set it up using `docker-compose`. Now, `docker-compose` is an official tool from Docker to define and stand up multi-container applications. The entire infrastructure will be configured in a YAML file, and then a single `docker-compose` command will do the rest of the job for you. More details are available at `https://github.com/apache/OpenWhisk-devtools/blob/master/docker-compose/README.md`.

- **Ansible**: OpenWhisk also supports deployment using Ansible, a popular configuration management tool. This is a good route if you want to do more customizations and have one or more VMs or machines at your disposal. Detailed installation instructions are available at `https://github.com/apache/OpenWhisk/blob/master/ansible/README.md`.

Out of all the methods, the preferred one for production deployment is Kubernetes followed by Ansible. Other options should be used only for development and experiment environments.

Now that we know about the deployment methods, let's proceed to set up a developer environment and try out OpenWhisk.

We will choose the QuickStart version of OpenWhisk for deployment as that is the easiest to start with. We will install the package and configure the OpenWhisk CLI (`wsk`) to manage the entities.

First, install the prerequisites: Docker, Java, NPM, and Node.js. I am using an Ubuntu 22.04 machine as my developer box here; please adjust your installation instructions based on the official documentation:

```
safeer@serverless103:~$sudo apt install docker.io
safeer@serverless103:~$ docker -v
Docker version 20.10.12, build 20.10.12-0ubuntu4
safeer@serverless103:~$sudo apt install default-jdk
safeer@serverless103:~$ javac --version
```

```
javac 11.0.15
safeer@serverless103:~$ sudo apt install nodejs
safeer@serverless103:~$ node -v
v12.22.9
safeer@serverless103:~$ sudo apt install npm
safeer@serverless103:~$ npm -v
8.5.1
```

Configure the JAVA_HOME environment variable:

```
safeer@serverless103:~$ ls -l /etc/alternatives/java
lrwxrwxrwx 1 root root 43 May  1 15:23 /etc/alternatives/java -> /usr/
lib/jvm/java-11-openjdk-amd64/bin/java
safeer@serverless103:~$ echo "JAVA_HOME=/usr/lib/jvm/java-11-openjdk-
amd64"|sudo tee -a /etc/environment
JAVA_HOME=/usr/lib/jvm/java-11-openjdk-amd64
safeer@serverless103:~$ source /etc/environment
safeer@serverless103:~$ echo $JAVA_HOME
/usr/lib/jvm/java-11-openjdk-amd64
```

Now, download and set up the QuickStart service:

```
safeer@serverless103:~$ git clone https://github.com/apache/openwhisk.
git
Cloning into 'OpenWhisk'...<OUTPUT TRUNCATED>......
safeer@serverless103:~$ cd openwhisk/
safeer@serverless103:~/openwhisk$ ./gradlew core:standalone:bootRun
```

The last step will throw a lot of output as it builds and sets up the standalone OpenWhisk service. Since this is running in the foreground, it is recommended that you run it in a screen or tmux session in parallel to your active shell. If all goes well, the whisk service will be started and the API will be listening on port 3233 and a playground UI will be on port 3232. Both the API and UI will be available only on the docker0 interface's IP. In the output, you will see a message similar to the following:

```
Launched service details

[ 3233 ] http://172.17.0.1:3233 (Controller)
[ 3232 ] http://172.17.0.1:3232/playground (Playground)

Local working directory - /home/safeer/.OpenWhisk/standalone/server-
3233
```

As you can see, the URLs can be found in the output of the services, but if you can't find them, first find the docker0 interface IP and then construct the URLs. Use the following command to extract the Docker IP (works on Linux) and test that both the API and the UI are working:

```
safeer@serverless103:~$ ip addr show dev docker0 scope global |awk '/
inet /{print $2}'|cut -d "/" -f1
172.17.0.1
safeer@serverless103:~$ curl -s http://172.17.0.1:3232
The requested resource temporarily resides under <a href="/playground/
ui/index.html">this URI</a>
safeer@serverless103:~$ curl -s http://172.17.0.1:3233|jq .api_paths
[
  "/api/v1"
]
```

Now that the service is running on the current terminal, let us switch to another terminal from where we will operate for the rest of this session. The first thing to do on the new shell is to install the OpenWhisk CLI. Download the CLI for your platform from `https://github.com/apache/OpenWhisk-cli/releases/`. Once you extract the `tar.gz` package, you will see the `wsk` binary. Move it to an appropriate location in the system path:

```
safeer@serverless103:~$ sudo mv wsk /usr/local/bin/
```

Now you have to configure the CLI. The `wsk` CLI needs two minimum parameters to work. The first one is the API endpoint, which in this case is `http://172.17.0.1:3233`, and the second is the auth token, which is predefined for the QuickStart as `23bc46b1-71f6-4ed5-8c54-816aa4f8c502:123zO3xZCLrMN6v2BKK1dXYFpX1PkccOFqm12CdAsMgRU4VrNZ9lyGVCGuMDGIwP`. Let us configure both the parameters from the CLI:

```
safeer@serverless103:~$ wsk property set    --apihost
'http://172.17.0.1:3233'    --auth '23bc46b1-71f6-4ed5-8c54-816aa4f8c50
2:123zO3xZCLrMN6v2BKK1dXYFpX1PkccOFqm12CdAsMgRU4VrNZ9lyGVCGuMDGIwP'
ok: whisk auth set. Run 'wsk property get --auth' to see the new
value.
ok: whisk API host set to http://172.17.0.1:3233
```

The CLI configuration is stored in a file named `.wskprops` in your home directory and will have the configuration parameters we set earlier. You can back up this file when playing around with CLI configuration. You can also list the properties/configuration using the CLI, as follows:

```
safeer@serverless103:~$ cat .wskprops
APIHOST=http://172.17.0.1:3233
AUTH=23bc46b1-71f6-4ed5-8c54-816aa4f8c502:123zO3xZCLrMN6v2BKK1dXYFpX1P
kccOFqm12CdAsMgRU4VrNZ9lyGVCGuMDGIwP

safeer@serverless103:~$ wsk property get
whisk API host        http://172.17.0.1:3233
whisk auth            23bc46b1-71f6-4ed5-8c54-816aa4f8c502:123zO3xZCLrM
N6v2BKK1dXYFpX1PkccOFqm12CdAsMgRU4VrNZ9lyGVCGuMDGIwP
whisk namespace       guest
client cert
```

```
Client key
whisk API version     v1
whisk CLI version     2021-04-01T23:49:54.523+0000
whisk API build       2022-04-06T01:58:03+0000
whisk API build number  3e3414c
```

Let us see whether the CLI is working by running some basic commands:

```
safeer@serverless103:~$ wsk list
Entities in namespace: default
packages
actions
triggers
rules
```

As you can see, the wsk CLI is configured and we are able to list the entities in the default namespace. Now let us start playing around with the OpenWhisk ecosystem.

Creating your first action

For any function to become an action, the following rules should be followed:

- Actions accept a dictionary as input and return another dictionary as output. The key-value pairs in the input should be JSON compatible.

- The function name (also referred to as the entry point to the action) should be named main. This can be changed by modifying the function definition and supplying a custom name for main.

Now let us create a simple action in Python for demonstration. We will start with a simple function that takes two numbers and returns the sum. The code is as follows:

```
safeer@serverless103:~/dev/pyfirst$ cat sum.py
def main(args):
    print(args)
    if 'first' not in args or 'second' not in args:
        return {'error': 'Not enough numbers to add'}
    elif not isinstance(args['first'],int) or
        not isinstance(args['second'],int):
        return {'error': 'arguments should be numbers'}
    else:
        return {'sum': args['first'] + args['second'] }
```

Now, create the action with the wsk CLI:

```
safeer@serverless103:~/dev/pyfirst$ wsk action create sum sum.py
ok: created action sum
safeer@serverless103:~/dev/pyfirst$ wsk action get sum
```

```
ok: got action sum
{
        "namespace": "guest",
        "name": "sum",
        "version": "0.0.1",
        "exec": {
        "kind": "python:3",
        "binary": false
        },
        "annotations": [
        {
                "key": "provide-api-key",
                "value": false
        },
        {
                "key": "exec",
                "value": "python:3"
        }
        ],
        "limits": {
        "timeout": 60000,
        "memory": 256,
        "logs": 10,
        "concurrency": 1
        },
        "publish": false,
        "updated": 1651465149420
}
```

Since the function is ready, let's prepare to invoke it by passing the arguments as a JSON file. Note that this is a synchronous (blocking) invocation (notice the -result argument to the wsk CLI):

```
safeer@serverless103:~/dev/pyfirst$ echo '{"first":10,"second":22}' >
sum.params.json
safeer@serverless103:~/dev/pyfirst$ wsk action invoke sum --param-file
sum.params.json  --result
{
        "sum": 32
}
```

Now let us invoke the same function asynchronously. If OpenWhisk accepts the invocation, it will return an activation ID, and we will use it to retrieve the action status:

```
safeer@serverless103:~/dev/pyfirst$ wsk action invoke sum --param-file
sum.params.json
ok: invoked /_/sum with id 4cc1c1177002492481c1177002c924a3
```

```
safeer@serverless103:~/dev/pyfirst$ wsk activation get 4cc1c1177002492
481c1177002c924a3
ok: got activation 4cc1c1177002492481c1177002c924a3
{
    "namespace": "guest",
    "name": "sum",
    "version": "0.0.1",
    "subject": "guest",
    "activationId": "4cc1c1177002492481c1177002c924a3",
    "start": 1651467113317,
    "end": 1651467113320,
    "duration": 3,
    "statusCode": 0,
    "response": {
    "status": "success",
    "statusCode": 0,
    "success": true,
    "result": {
        "sum": 32
    }
    },
    "logs": [
    "2022-05-02T04:51:53.319772813Z {'first': 10, 'second': 22}"
    ],
    "annotations": [
        {
            "key": "path",
            "value": "guest/sum"
        },
    },
<OUTPUT TRUNCATED>
```

As you can see, the function was asynchronously executed and we retrieved the result using the activation ID that was returned.

Triggers and rules

Now that we understand the basics of creating actions, let's try to create a trigger and associate it with this action using a rule:

```
safeer@serverless103:~/dev/pyfirst$ wsk  trigger create basicMath
ok: created trigger basicMath
safeer@serverless103:~/dev/pyfirst$ wsk trigger get basicMath
ok: got trigger basicMath
{
```

```
    "namespace": "guest",
    "name": "basicMath",
    "version": "0.0.1",
     "limits": {},
     "publish": false,
     "rules": {
    "guest/sumRule": {
        "action": {
                "name": "sum",
                "path": "guest"
            },
            "status": "active"
        }
    },
    "updated": 1651483435409
}
```

Next, associate this with our sum function using a rule:

```
safeer@serverless103:~/dev/pyfirst$ wsk rule create sumRule basicMath
sum
ok: created rule sumRule
safeer@serverless103:~/dev/pyfirst$ wsk rule get sumRule
ok: got rule sumRule
{
        "namespace": "guest",
        "name": "sumRule",
        "version": "0.0.1",
        "status": "active",
        "trigger": {
        "name": "basicMath",
        "path": "guest"
        },
        "action": {
        "name": "sum",
        "path": "guest"
        },
        "publish": false,
        "updated": 1651483589056
}
```

Now this rule has associated the trigger with our action. The action will run every time the trigger is fired. Triggers are fired when it receives a dictionary of key-value pairs. This is an event and it can be fed to the trigger either by a feed or manually using the `wsk` CLI/API. Since we haven't created a feed, let's do this manually and pass the trigger an event that the `sum` action will understand – a dictionary of two numbers. We will use the input from the `sum.params.json` parameters file we used with the `sum` action earlier:

```
safeer@serverless103:~/dev/pyfirst$ wsk trigger fire basicMath
--param-file sum.params.json
ok: triggered /_/basicMath with id ff34e8e5ad414731b4e8e5ad412731f7

safeer@serverless103:~/dev/pyfirst$ wsk activation get --last
ok: got activation b02c245ac3504b49ac245ac350ab4900
<OUTPUT TRUNCATED>
```

In the preceding output, you can see the same result that you saw earlier when we manually invoked the action.

If you just want to see the output of an activation and not the rest of the details, you can do the following instead:

```
safeer@serverless103:~/dev/pyfirst$ wsk activation result
b02c245ac3504b49ac245ac350ab4900
{
      "sum": 32
}
```

Next, let us look at creating packages.

Packages

Next, let us explore packages. As we discussed earlier, a package is a way to bundle together a bunch of related actions in a shareable manner. It can contain both actions and feeds. A package cannot, however, contain another package.

Let's start by creating a simple package:

```
safeer@serverless103:~/dev/pyfirst$ wsk package create mathops
ok: created package mathops
safeer@serverless103:~/dev/pyfirst$ wsk package get mathops
ok: got package mathops
{
      "namespace": "guest",
      "name": "mathops",
      "version": "0.0.1",
      "publish": false,
```

```
        "binding": {},
        "updated": 1651504484929
}
```

Now let us add an action to this package. We will re-use the sum.py action for this:

```
safeer@serverless103:~/dev/pyfirst$ wsk action create mathops/sum sum.
py
ok: created action mathops/sum

safeer@serverless103:~/dev/pyfirst$ wsk package get --summary mathops
package /guest/mathops
   (parameters: none defined)
 action /guest/mathops/sum
   (parameters: none defined)
```

Now let's invoke the function that is inside the package:

```
safeer@serverless103:~/dev/pyfirst$ wsk action invoke --result
mathops/sum --param-file sum.params.json
{
      "sum": 32
}
```

Default parameters can be set on a package, and all entities within the package will have access to them. Let's set the default values for the sum action and see how it works:

```
safeer@serverless103:~/dev/pyfirst$ wsk package update mathops --param
first 5 --param second 7
ok: updated package mathops
```

Now let us see whether the default values are accepted when we invoke the action without parameters:

```
safeer@serverless103:~/dev/pyfirst$ wsk action invoke --result
mathops/sum
{
      "sum": 12
}
safeer@serverless103:~/dev/pyfirst$ wsk action invoke --result
mathops/sum --param first 10
{
      "sum": 17
}
```

As you can see, the default values are applied and can be overridden by parameters provided at invocation time. Let us now publish this package so that it is available across namespaces:

```
safeer@serverless103:~/dev/pyfirst$ wsk package update mathops
--shared yes
ok: updated package mathops
```

Now this package will be available for use from any namespace. We will explore feeds next.

Feeds

OpenWhisk allows users to create and expose an event producer interface as a feed within a package. Usually, there are pre-built feeds that are open source or are provided by cloud vendors such as IBM. Designing and creating your own feed is usually an advanced use case and most normal users won't need to create one. We will touch upon feed architecture a little bit and then move on to a more common use case – **web actions**.

There are three types of feed architectures. They are categorized as follows:

- **Hooks**: In this case, the event producer would support webhooks and will post to the web endpoint exposed by the feed whenever an event is available. This is ideal for low-volume event streams.

- **Polling**: As the name indicates, in the polling architecture pattern, the feed action will keep polling the event producer endpoint periodically for any new events. While this is a comparatively easy implementation, the availability of new events depends on the polling interval and can be an issue for low-latency applications.

- **Connections**: In this mode, any service that runs always needs to be configured to maintain a persistent connection with the event source.

A feed action is like a normal action but it has to handle some of the selected lifecycle events (of an event) and other parameters, such as `triggerName` and the auth key.

To find out more about feeds and their implementation, visit the official documentation at `https://github.com/apache/OpenWhisk/blob/master/docs/feeds.md`.

Web actions

Web actions are regular actions that are *marked* to be exposed as an API endpoint. This marking is called annotations. Annotations are applicable to all OpenWhisk assets – actions, packages, triggers, and rules. They are similar to function annotations and add a qualifier/feature enhancement to the action. They are attached to resources as a key (which is the name of the annotation) and the value (the definition or value of the annotation). What can be annotated on an asset is very loosely defined, and you can add your own annotation keys. While that is the case, most keys are used for documenting the asset itself, and some of them are used for enabling features such as web APIs. Let us look at adding a simple description as an annotation:

```
safeer@serverless103:~/dev/pyfirst$ wsk action update mathops/sum -a
'description' 'Function to find the sum of two numbers'
ok: updated action mathops/sum
```

Let us retrieve and examine the annotations of this action:

```
safeer@serverless103:~/dev/pyfirst$ wsk action get mathops/sum|awk
'{if( NR != 1 ) print}'|jq .annotations
[
  {
      "key": "provide-api-key",
      "value": false
  },
  {
    "key": "exec",
    "value": "python:3"
  },
  {
      "key": "description",
      "value": "Function to find the sum of two numbers"
  }
]
```

As you can see, our annotation is added to the action, and there are already two annotations that were added by default when the function was created. To find out more about annotations, visit https://github.com/apache/OpenWhisk/blob/master/docs/annotations.md.

Now that we understand annotations, let us see how to create web actions using annotations. We will re-use the same sum function – but with necessary modifications to suit the web – for this:

```
safeer@serverless103:~/dev/pyfirst$ cat sum-web.py
import json
def main(args):
      first = int(args['first'])
      second = int(args['second'])
      return { 'statusCode': 200, 'body': {'sum': first + second } ,
          'headers': { 'Content-Type': 'application/json' }
            }
```

Note that the return value is modified to fit that of an HTTP response. Also, I have trimmed down the error checking to keep the code short. Now let us use this code to create a web action:

```
safeer@serverless103:~/dev/pyfirst$ wsk action create sum-web sum-web.
py --web true
```

```
ok: created action sum-web

safeer@serverless103:~/dev/pyfirst$ wsk action get sum-web  --url
ok: got action sum-web
http://172.17.0.1:3233/api/v1/web/guest/default/sum-web

safeer@serverless103:~/dev/pyfirst$ curl  "http://172.17.0.1:3233/api/
v1/web/guest/default/sum-web?first=6&second=20"
{
   "sum": 26
}
```

As you can see, the sum action is working well over HTTP. You can pass headers, statusCode, and body into the HTTP response from the action. As for input, parameters are passed to the action input argument (the dictionary that is passed to the action) as its keys. The HTTP context that carries relevant request information is passed to the actions as specially named keys of the argument. Each of the items in the HTTP context is passed as keys prefixed with __ow_<parameter_name>. Some of those parameters are __ow_method, __ow_headers, __ow_path, __ow_body, __ow_user, and __ow_query. Most of these parameters are self-explanatory.

You can enable authentication for the endpoint using the -web-secure argument of the wsk CLI or by annotating the function with the require-whisk-auth key. You can either define the authentication token or use OpenWhisk identities for this auth method. If this doesn't suit your need, you are free to implement authentication and authorization in your actions.

There is a lot more to web actions than what we discussed. To find out more, please visit https://github.com/apache/OpenWhisk/blob/master/docs/webactions.md.

While web actions are the simple way to manage an endpoint, OpenWhisk provides an advanced API management solution. This is the API gateway and it can act as a proxy between users and web actions, providing custom routing, authentications, tenant management, and so on. This API gateway is implemented on top of OpenResty – a custom version of Nginx with Lua support. It also needs Redis for its backend. Covering the API gateway in detail is outside the scope of this chapter. Please refer to the official documentation for more details: https://github.com/apache/OpenWhisk-apigateway.

Administration and deployment

OpenWhisk exposes a number of features for the effective administration of the clusters and their entities. This is intended for the operators of the cluster and not the developers. In order to manage OpenWhisk, you need to install and set up the wskadmin CLI. Please refer to the documentation on this at https://github.com/apache/OpenWhisk/blob/master/tools/admin/README.md.

When it comes to deployment, so far we have deployed actions and created packages, rules, and triggers using the CLI. There is another way to do these at scale and in a more structured way. It allows you to describe your OpenWhisk entities using a manifest file written in YAML format. This helps developers to define OpenWhisk entities as code/configuration and manage them via git. There is another CLI – **wskdeploy** – that should be used to apply the manifest file against an OpenWhisk installation. It also introduces the concept of projects to collectively manage a set of entities. Refer to the documentation here: `https://github.com/apache/OpenWhisk-wskdeploy`.

Secret management is a common requirement for any application deployment service, including FaaS. While OpenWhisk doesn't provide any secret management capabilities, you can pass the credentials as parameters by updating the action (preferably using a parameter file). These are called default parameters. Refer to the parameter documentation to find out more: `https://github.com/apache/OpenWhisk/blob/master/docs/parameters.md`.

Project – IoT and event processing with IBM Cloud functions

The purpose of this project is to collect temperature information sent by an IoT sensor, persistently store it for future analysis, and also send out near-real-time alerts when the temperature crosses a specific threshold.

IoT devices send their information to IoT platforms using the MQTT protocol. Most IoT clouds will provide a gateway that acts as an MQTT broker and receive these messages. Devices publish their sensor information to the MQTT topic and then the data can be consumed by various applications for business processing.

The IoT platform provided by IBM Cloud is called the IBM Watson IoT Platform. Once you provision an instance of this platform on the cloud, you can use the details to obtain the MQTT gateway address as well as add a device to the platform. This device will represent our pseudo-IoT device, which will be a heat sensor that will send out the current timestamp and temperature at defined intervals. We will use a Python script that can send MQTT messages instead of a real IoT device, as most readers won't have one handy. Also, it doesn't matter for the demonstration of OpenWhisk/IBM Cloud function capabilities.

The compute workhorse of this project will be the IBM Cloud functions, which are the managed OpenWhisk offering from IBM Cloud. We will use multiple IBM Cloud functions (OpenWhisk actions) for different parts of the processing. The first function will subscribe to the MQTT topic and read the sensor data sent by the device, and then insert it into a Cloudant NoSQL database (a managed Couchbase service offered by IBM Cloud). Cloudant provides an OpenWhisk feed for event processing, which can be used to fire a trigger when any change happens in a Cloudant database. The trigger will associate with a sequence of IBM Cloud functions to process the records added to the database. In our case, we want an alert if the temperature goes above a certain threshold.

The following is the architecture diagram of this project:

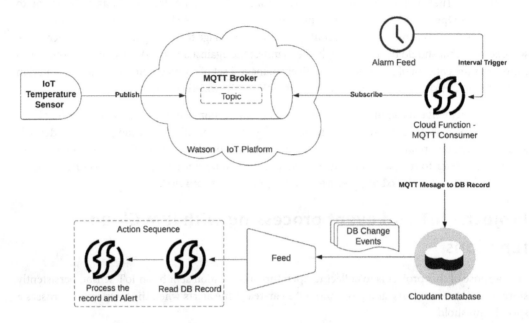

Figure 8.6 – Project architecture

Let us look at the execution flow in detail.

The IoT sensor is a Python script that emulates the behavior of an IoT device by sending temperature information as MQTT messages to an MQTT topic provided by the IBM Watson IoT platform.

Since MQTT is a message queue, these events can be processed by a consumer asynchronously. For this, we will use an IBM Cloud function (a hosted OpenWhisk action). In a full-fledged and large-scale IoT environment, you will need a consumer that runs as a service full time. But given that we are testing a small use case and we can consume the messages asynchronously, we would rather use a IBM Cloud function that will be scheduled to run once every minute to process available messages and then exit. This will also help you to familiarize yourself with a special set of feeds provided by IBM Cloud functions to schedule tasks.

To run a function on a schedule, we need to use a feed that is provided by IBM. They provide a package called **alarm**, which has a number of feeds. One of them is the interval feed, which can be used to trigger a function at defined intervals. We will use this feed to trigger the consumer function every minute. Every time this function is launched, it will subscribe to the MQTT topic and pull all pending messages. These messages will then be written to a database in an IBM NoSQL database called cloudant.

The preceding steps complete the first part of the requirement – that is, persistently storing the sensor data. Now we need to use event processing to process messages and alert when a threshold breach occurs. In a small setup, you can do this within the subscriber function that reads the MQTT message. But given that our idea is to demonstrate event processing using feeds and actions, let us do it in a slightly more complicated way.

The Cloudant database provides an OpenWhisk feed to the IBM Cloud functions. This is part of the Cloudant OpenWhisk package, which also provides utility functions for reading and writing into the database. The feed allows users to trigger actions whenever a record in the Cloudant database changes. We will use this for our event processing mechanism.

When we set up this Cloudant feed and the trigger is fired, it will send out the ID of the changed record, among other details of the event. This can be used to retrieve the database record and process it. We will chain two functions (an action sequence) for this. The first is the read function provided by the Cloudant package, which will take the record ID from the feed and retrieve the record, which will be passed down to the second function in the sequence. This function will process the record and alert based on the temperature threshold.

This completes the second requirement of the project to alert based on the threshold.

More details and implementation instructions are provided in the GitHub repo at `https://github.com/PacktPublishing/Architecting-Cloud-Native-Serverless-Solutions/tree/main/chapter-8`.

Summary

While OpenWhisk is envisioned as a vendor-neutral FaaS platform, multiple vendors have adopted it as their FaaS platform. The prominent vendors are IBM and Adobe. IBM exposes most of the OpenWhisk features and runtime support without restrictions through its IBM Cloud functions. Adobe has a more customized and restricted version of OpenWhisk on its **Adobe I/O Runtime** platform. I/O Runtime allows only JavaScript on the Node.js runtime and runs on Adobe's Experience platform.

In this chapter, we covered the programming model of OpenWhisk followed by its architecture and deployment. We also created a sample action along with triggers and rules. We learned about web actions, feeds, the API gateway, and more. This is a great platform with a lot of potential, and businesses that want to take control of their FaaS environment can definitely consider OpenFaaS as an ideal choice.

This concludes our chapter on OpenWhisk as well as the second part of this book, where we covered a number of serverless platforms and vendors. In the next part of the book, we will look into programming frameworks, best practices, and a lot more.

Part 3 – Design, Build, and Operate Serverless

Developing serverless is only the first part of the serverless journey. In this section, we will look into developing tools and frameworks that make it easy to build serverless applications and provision resources. Then, we will look into necessary security practices and recommendations for serverless. We will wind up this part and the book by covering architectural patterns for serverless and serverless design for some of the common architecture designs.

This part has the following chapters:

- *Chapter 9, Implementing DevOps Practices for Serverless*
- *Chapter 10, Serverless Security, Observability, and Best Practices*
- *Chapter 11, Architectural and Design Patterns for Serverless*

9

Implementing DevOps Practices for Serverless

In the previous section, we went through several serverless platforms and technologies. This should give you a fair idea of how diverse the serverless landscape is. Now, while these technologies might be diverse and your code and infrastructure for these platforms could be in any language or framework of your choosing, at the end of the day, they all have the same challenges in getting deployed into production. Serverless infrastructure and FaaS code ultimately follow the same development cycle as any other software or framework. They face the same challenges of developing, testing, and deploying code as well.

Similar to what happens in the usual software ecosystem, serverless can also become complicated and multi-platform. As time goes by, it can be clearly seen that using just the vendor tools to manage the serverless stack is not enough. This gets even more complicated when serverless stacks from multiple vendors are used. This is where serverless frameworks came into the picture. There are broadly two sets of frameworks that serverless stacks use. The primary category contains the frameworks that are aimed at automating the serverless-specific services of the cloud vendors. Then comes the generic frameworks that fall into the category of **infrastructure as code (IaC)**. These frameworks can be used to automate all infrastructure and service pieces of the cloud, including serverless.

In this chapter, we will explore the following:

- The serverless framework
- The Zappa framework
- Infrastructure as code with Terraform
- Infrastructure as code with the Pulumi SDK
- CI/CD pipelines for serverless
- Testing serverless

In the coming section, we will briefly look into some aspects of the serverless **software development life cycle (SDLC)**.

General SDLC practices for serverless

While serverless differs from normal software applications in the execution model, the development and deployment process is similar. Unlike a full-fledged software application, serverless functions are much more fragmented and perform atomic business functions. If a good SDLC practice is not followed, they could end up like the runaway scripts that you had tucked into the corners of your server infrastructure in the old times, hard to troubleshoot and even harder to find when things break unexpectedly.

With the scattered nature of FaaS code and other serverless services, it is critical to have a well-designed architecture. The architecture and functions should be well documented to enable new developers to understand the system better. Learning by reverse engineering is not very easy in the serverless world. Ensure that all your code is documented and that the serverless infrastructure that supports these functions is managed using IaC. This will allow you to better document your infrastructure along with business logic. Language support is another thing to consider before you start with development. While most FaaS platforms support popular languages, there are always some languages or versions that are not supported (yet) by a cloud platform.

The development in itself follows the usual application development practices. You should decide on a Git workflow that works in your organization. You could use workflows such as trunk-based development or feature/branch-based development. While the functions developed are small, ensure that you have a good peer review and pull request-based development practice. The natural next steps of the development cycle are test, build, and release followed by deployments to different environments. A good CI/CD pipeline is critical for these stages. We are going to cover some of these important aspects in the last two sections of this chapter.

In the next section, we are going to cover one of the most popular serverless frameworks – called Serverless itself!

The serverless framework

The *serverless framework* is precisely what the name suggests: a framework for developing and deploying serverless applications, and arguably the most popular one at that. This open source framework was created in 2015 and was initially called **JavaScript Amazon Web Services (JAWS)**. In 2016, the serverless framework team rebooted their code with major changes and renamed it Serverless. They established Serverless Inc to drive the project as well as to provide commercial support and value-added services on top of the framework. The initial focus of the framework was on AWS serverless – understandably so since they were the primary players in the field and other cloud vendors were playing catch up. But they later extended the framework to support the serverless platforms of more cloud vendors.

At present, serverless is a framework and tool that enables developers to create serverless applications and deploy them across most of the popular serverless vendors. The framework itself is written in JavaScript but doesn't limit writing your functions in the same language. It supports many languages for each vendor and keeps adding more support. In addition to the framework and developer tools, it also provides CI/CD capabilities and monitoring. The framework can also be extended using plugins to add more capabilities for managing the development and deployment process. The serverless team have also introduced a serverless cloud of their own infrastructure backed by AWS (in public preview right now) and a *serverless console* for monitoring your serverless applications.

The most important advantage of using the serverless framework is the ability to use common constructs across any cloud. This helps developers in defining and developing serverless applications without knowing the vendor platform in depth. This increases the developers' velocity, improving the overall time to market. The framework and tools are open source, which helps with faster adoption. Even the serverless cloud offers a free tier for single developers. Another advantage is that it takes care of configuring the right permissions and creating the right cloud resources, helping to avoid misconfigurations and security loopholes. The ability to extend the framework with plugins gives more control to developers who need extra capabilities or custom integrations. It also offers boilerplate code and templates, thus reducing the barrier to entry further.

Now that we have an understanding of what the serverless framework has to offer, let's go ahead and experiment with the framework a little bit.

Getting started with the serverless framework

The serverless framework has two major parts – the serverless CLI and the hosted dashboard. The hosted dashboard is free and you can use it to deploy your serverless apps very quickly. The CLI in itself is enough to develop the serverless application using the framework, but having the dashboard helps in the full application life cycle management, including monitoring, troubleshooting, and the CI/CD pipeline.

We will start by using the CLI and later explore the dashboard. The CLI itself is written in JavaScript and you will need the npm package manager to install it. Also, we will be using AWS as the cloud of choice to deploy our applications – though it wouldn't matter much which cloud we used, given serverless abstracts away most of the AWS complexities. However, note that some of the concepts we cover will be specific to the cloud vendor and there may be changes in how these concepts are implemented/configured for a different vendor. So, for any vendor other than AWS, be sure to cross-reference the documentation.

Let us start by installing the CLI itself. If you don't have npm installed, you will need to install both Node.js and npm. Refer to the following docs for more details – `https://docs.npmjs.com/downloading-and-installing-node-js-and-npm`.

Once npm is installed, you can install the serverless CLI, as follows:

```
npm install -g serverless
```

This will provide you with a CLI named `serverless`.

It is also possible to install the CLI as a standalone binary, but that is not recommended, and if possible, stick to the npm-based installation.

The next step is to configure your cloud credentials (in this case, AWS). As a prerequisite, you should create an AWS IAM account with sufficient privileges and obtain the access and secret keys for the account. Follow the instructions at the following link to do so: `https://docs.aws.amazon.com/general/latest/gr/aws-sec-cred-types.html`.

Once you obtain the credentials, install the AWS CLI and configure it to use these credentials. The following two links have the instructions to do so:

- Installing the AWS CLI – `https://docs.aws.amazon.com/cli/latest/userguide/getting-started-install.html`

- Setting up the AWS CLI – `https://docs.aws.amazon.com/cli/latest/userguide/getting-started-quickstart.html`

The next step is to sign up for a free account with `serverless.com` to access its serverless dashboard. This is an optional step but will be helpful in monitoring and troubleshooting.

To sign up, visit `https://app.serverless.com`. Once you sign up, you can visit `https://app.serverless.com/<your_username>/settings/providers` to configure your AWS credentials so that the dashboard can pull information from your cloud account. Please note that at this time, the dashboard only supports AWS as a provider with language support for node and Python in a selected set of regions. This limitation doesn't apply to the framework and CLI tool.

At this point, you are all set to get used to the CLI and start creating your first serverless application.

To start developing with the serverless framework, first, you need to create a serverless service. A service is the unit of organization of the framework. Consider this similar to a self-contained project that consists of one or more functions along with its code, config, and associated infrastructure definitions. When creating a project, you need to provide a template as input. A template defines a barebones structure to create the serverless project.

Now, let us start by creating a project. I am choosing `aws-python3` as the template. Let us run the following command to create the project with the given template in a directory called `projectpy3`:

```
$serverless create --template aws-python3 --path projectpy3
```

Let us see what is inside the project directory:

```
$tree projectpy3
projectpy3
├── handler.py
└── serverless.yml
```

```
0 directories, 2 files
```

As you can see, there are two files in the directory, `serverless.yml` and `handler.py`. Let us understand what both of them do.

`serverless.yml` is the configuration file for the serverless framework. Most of the magic happens here. It contains all the information required to create the project with function code, configuration, as well as the associated infrastructure. We will discuss some of the key elements of this file in the coming sections. One thing to remember about `serverless.yml` is that it can also support different formats than YAML. The currently supported formats are JSON (`.json`), JavaScript (`.js`), and TypeScript (`.ts`).

`handler.py` is the file that will contain the function definition for one or more FaaS functions. In this case, it will be AWS Lambda in Python. If the code were written in JavaScript, the filename would be `handler.js`. This file and the functions within it are referenced in the `serverless.yml` file.

Before we dive further into the `serverless.yml` file, let us understand the core concepts of serverless frameworks.

Framework concepts

The core purpose of the framework is to make it easy for developers to build event-driven serverless architecture-based applications while the framework abstracts out the necessary infrastructure management. For AWS providers, the main concepts of the framework are covered next.

Functions

This is the function code that the cloud platform will execute on their FaaS platform, for example, Lambda. As discussed in *Part 1* of the book, a function is a unit of execution that caters to a particular business feature or a single job. Functions are a root-level property in the `serverless.yml` file. You can define multiple functions under the `functions` root property. The following is a sample function block extracted from a `serverless.yml` file:

```
functions
  hello:
    handler: handler.hello
    name: my-lambda
    description: Greets the caller
    runtime: python3.8
```

Most of these properties are self-explanatory. `handler` is the function name that is defined in `handler.py`, `name` is the name of the function, `description` is the description of the function, and `runtime` is the language runtime in which the function will be run. Some of the other optional properties are `memorySize`, `timeout`, `environment`, `provisionedConcurrency`, reservedConcurrency, and tracing. Some of these are specific to the vendor (AWS in this case), while some others are generic.

Events

Events trigger functions and can come from other cloud resources or as a direct invocation by the developer. Some of the common triggers in AWS are as follows:

- HTTP request to API Gateway URL

- S3 bucket object upload

- SNS message

- CloudWatch/EventBridge schedule

Serverless will automatically provision any cloud infrastructure that might be needed to produce the event. All you need to do is define the event in the serverless.yml file. Events are a sub-property of the function root property. A sample event definition for the API Gateway is as follows:

```
functions:
  hello:
    handler: handler.hello
      events:
          - httpApi:
              path: /
              method: get
```

Each type of event has its own properties and will be specific to the event type. As you can see from the sample given, an event is an array element, which means you can define multiple events as triggers of the same function as long as the vendor supports it.

Resources

Resources are cloud infrastructure components that are needed by the function. For AWS Lambda, these could be AWS services such as S3 buckets, API Gateway, SQS queues, and SNS topics.

In the context of AWS, the resources are deployed as a cloudformation stack. If you remember our discussion from *Chapter 3*, AWS CloudFormation is an IaC service provided by AWS itself to automate infrastructure provisioning. In serverless.yml, resources is a root-level property and the definition in itself is a cloudformation template (which is AWS services defined as code). The following is a sample from serverless.yml defining an S3 bucket:

```
resources:
  Resources:
    S3Bucket:
```

```
Type: 'AWS::S3::Bucket'
Properties:
  BucketName: album-image-storage
```

Services

A service is the framework's basic unit of organization, as discussed before. Consider it like a project that you will use for one business function. A service is defined by just one root-level property, `service`, which is the name of the service itself. One `serverless.yml` file corresponds to one serverless service. If you have a complicated infrastructure with multiple serverless services, you can use a `serverless-compose.yml` file to tie them all together and deploy them together.

Another root property that is tied to the service is the vendor. The `vendor` property defines the cloud provider that is backing this service. It has sub-properties that will define vendor-specific settings. An example would be the default region where the functions should be deployed. You can also have global overrides for other properties, for example, the runtime can be overridden globally at the vendor level if most of your functions are going to use the same runtime.

The following is a sample service and vendor definition:

```
service:
  projectpy3
provider:
  name: aws
  runtime: python3.8
  region: ap-south-1
  memorySize: 256
```

Plugins

Plugins can override or extend the functionality of the framework, as necessary. There is a root property in `serverless.yml` where you can define which of the available plugins you want to enable for this service. You can use the serverless CLI to install these plugins in your local environment. For example, there is a plugin to manage third-party package requirements for services written in Python. You can install this plugin using the following command:

```
serverless plugin install -n serverless-python-requirements
```

This covers the main concepts – some of which are common across vendors while some are specific to AWS. Some of the other vendors have other fundamental building blocks than what we have covered here. Please refer to the documentation for those. In the next section, we will see how to update and deploy the service we built in the previous section.

Updating and deploying the serverless service

Now, let us go back and look at how to modify and deploy our project. We are going to start with a simple function that will return the location details of an IP address when that IP address is provided as input. Behind the scenes, we will use the open API from ip-api.com to obtain this data. Let us look at the serverless.yml and handler.py files for this. The code and config are self-explanatory, so we won't get into the details:

serverless.yml

```
service: projectpy3
frameworkVersion: '3'
provider:
  name: aws
  runtime: python3.8
functions:
  ipinfo:
    handler: handler.ipinfo
```

Handler.py

```
import json,requests
def ipinfo(event, context):
    ip = event['ip']
    body = ipapi(ip)
    return { "statusCode": 200, "body": body }
def ipapi(ipaddr):
    response = requests.get("http://ip-api.com/json/"+str(ipaddr))
    return response.json()
```

> **Note**
> Note that the code doesn't contain any error checking or input validation. This is to keep the content size in check. Be sure to add necessary validations to your code.

Before we deploy the code to AWS, we can invoke the function locally to ensure the code is working fine. If everything works, the invocation will return a JSON object. To parse it and return only the country name from ipinfo, we will need to install the jq CLI. Refer to https://github.com/stedolan/jq for installation and usage instructions. Switch to the project/service directory first and run the following command:

```
$serverless invoke local --function ipinfo --data
'{"ip":"8.8.8.8"}'|jq .body.country
"United States"
```

As you can see, the local invocation worked fine and returned the right result. Before we deploy this function, we need to install the Python package dependency of this function – the `requests` package is not available by default in Lambda and needs to be packaged along with our function. Follow the given steps to do this. Docker installation is a prerequisite for this process:

```
$python3 -m venv venv
$source venv/bin/activate
$pip install requests
```

For the following command, accept all the defaults:

```
$npm init
$npm install --save serverless-python-requirements
```

Add the following root-level properties to `serverless.yml`:

```
plugins:
  - serverless-python-requirements
custom:
  pythonRequirements:
    dockerizePip: non-linux
```

If you want to understand in detail why these steps are needed, refer to `https://www.serverless.com/blog/serverless-python-packaging/`. Now you can deploy the function as follows:

```
$ serverless deploy
...OUTPUT TRUNCATED...
functions:
  ipinfo: projectpy3-dev-ipinfo (827 B)
```

You can see that your function is deployed as Lambda, as follows:

```
$aws lambda list-functions --out json|jq ".Functions[].FunctionName"
"projectpy3-dev-ipinfo"
```

Now, let us associate this with API Gateway as an event source to invoke the function as an API.

The relevant part of the `serverless.yml` file is as follows:

```
functions:
  ipinfo:
    handler: handler.ipinfo
    events:
      - httpApi:
          method: GET
          path: /ipinfo/{ip}
```

Given that the event that the API Gateway will send will be different from what we use during our direct invocation, let us make the following change in the handler:

```
def ipinfo(event, context):
    ip = event['pathParameters']['ip']
    body = ipapi(ip)
    return { "statusCode": 200, "body": json.dumps(body) }
```

Now, deploy the function as follows:

```
root@serverless104:~/projectpy3# serverless deploy
...OUTPUT TRUNCATED...
endpoint: GET - https://93bavu44t6.execute-api.us-east-1.amazonaws.
com/ipinfo/{ip}
...OUTPUT TRUNCATED...
```

As you can see, we got an API Gateway endpoint in return. Let's hit it with the IP as a parameter:

```
$curl -s https://93bavu44t6.execute-api.us-east-1.amazonaws.com/
ipinfo/8.8.8.8 |jq .country
"United States
```

As you can see, this returns the IP information similar to how we invoked the function locally. Once you have finished your experiments, be sure to clean up all the cloud resources you created using the following:

```
serverless remove
```

Other features of serverless in a nutshell

There is a lot more to this framework than we covered here. Some of the key features are as follows:

- **Event sources**: Serverless supports all event sources that AWS Lambda itself supports.
- **Parameters**: Parameters allow you to customize your deployment by parameterizing key configurations. They can also be used for storing secrets.
- **Variables**: Variables allow users to dynamically replace config values in the `serverless.yml` config.

We have covered the basics of serverless frameworks here. Be sure to check out the official documentation for up-to-date information. Note that the framework has gone through multiple iterations and the current version is 3. In the next section, we will look at another framework for serverless, Zappa, which is focused on Python.

Zappa – the serverless framework for Python

The serverless framework started as a way to easily deploy serverless applications in AWS and then expanded to support a large number of cloud vendors. Within each vendor, it also support several languages. Serverless is trying to become the most generic and popular serverless framework by supporting every popular cloud and language combination. Zappa goes against this philosophy and is purpose-built for a smaller set of use cases but does that well.

Zappa is a framework and tool that makes it easy to deploy serverless and event-driven applications written in Python on AWS. It was created by the people behind **Gun.io**. Zappa can convert existing Python web applications into serverless by integrating them with AWS Lambda and API Gateway. Zappa also supports event-driven serverless applications where your Python code can be executed against events generated in AWS. You can also schedule your Python functions – after Zappa converts them into Lambda functions.

Given that this framework is specific to Python and AWS, some familiarity with these two would be good for this section, but the examples are made extremely simple to support those of you who may not be familiar with either of these.

Zappa supports web applications, event-driven applications, and scheduled functions. In this section, we will use a web application to demonstrate the capabilities of the framework. We will reuse the same IP information API code that we used in the previous section. Given that Zappa's expertise is in converting existing Python code into serverless, we will first create a Python web application that provides the same API that we created in the previous section. We will use a Python framework called **Flask** for this. Flask is a web application framework that falls into the category of **Web Server Gateway Interface (WSGI)**. WSGI is a Python standard that specifies how a web server communicates with Python web applications or frameworks and how they can be chained together to process HTTP requests. Zappa supports all WSGI-compatible Python web applications. We are choosing Flask as it is the simplest framework to create and deploy Python web applications.

Creating and testing the IP information API in Flask

The core logic doesn't change, but we need to create a Flask project, install the necessary dependencies, and then test the code. First, let us set up the project. Please note that at the time of writing this book, 3.9 is the newest version of Python that is supported by AWS, and subsequently by Zappa. Ensure that you have the right installation.

Create the project directory and switch to it:

```
mkdir ~/zappa-ipinfo
cd ~/zappa-ipinfo
```

Create a virtual environment and install the necessary dependencies:

```
python3.9 -m venv
source venv/bin/activate
pip install requests flask
pip freeze > requirements.txt
```

Now, let us create the Flask code. Open a file named app.py and add the following code:

```
from flask import Flask
import json
import requests

app = Flask(__name__)

@app.route('/ipinfo/<ip>')
def getinfo(ip):
    body=ipapi(ip)
    return body

def ipapi(ipaddr):
    response = requests.get("http://ip-api.com/json/"+str(ipaddr))
    return response.json()

if __name__ == "__main__":
  app.run()
```

As you can see, the code uses the same ipapi function from the previous section. Then, we define a parameterized route (API endpoint) and bind it to a getinfo function, which calls the ipapi function with the provided IP parameters. Now, let us use Flask's built-in capability to run a test server and hit the API endpoint to locally test the code:

```
$python3 app.py
WARNING: This is a development server. Do not use it in a production
deployment. Use a production WSGI server instead.
 * Running on http://127.0.0.1:5000
```

Now, let us check the API:

```
curl -s http://localhost:5000/ipinfo/8.8.8.8|jq .country
"United States"
```

As you can see, the API is working as expected. Now, let us use Zappa to convert this into an AWS serverless application. First, you need to install the Zappa Python package in the same virtual environment as the application, which will provide us with a Zappa CLI:

```
pip install zappa
```

Now, you can initialize the Flask project as a Zappa project using the `zappa init` command. This will launch an interactive session where Zappa will collect some basic information and create a `zappa_settings.json` file. This file is the metadata for further Zappa CLI actions. A lot of verbose output is truncated in the following output to save space. Remember that you should have your AWS IAM account and AWS CLI set up for that to work. See as follows how Zappa project initialization works:

```
(venv) root@serverless104:~/zappa-ipinfo# zappa init


Your Zappa configuration can support multiple production stages, like
'dev', 'staging', and 'production'.
What do you want to call this environment (default 'dev'): dev


AWS Lambda and API Gateway are only available in certain regions.
Let's check to make sure you have a profile set up in one that will
work.
Okay, using profile default!


Your Zappa deployments will need to be uploaded to a private S3
bucket.
What do you want to call your bucket? (default 'zappa-o6ssrdxvn'):
zappa-flask-test-abcd


It looks like this is a Flask application.
Where is your app's function? (default 'app.app'): app.app


Would you like to deploy this application globally? (default 'n')
[y/n/(p)rimary]: n


Okay, here's your zappa_settings.json:


{
    "dev": {
    "app_function": "app.app",
    "aws_region": "ap-south-1",
    "profile_name": "default",
    "project_name": "zappa-ipinfo",
    "runtime": "python3.9",
    "s3_bucket": "zappa-flask-test-abcd"
```

```
        }
}

Does this look okay? (default 'y') [y/n]: y

Done! Now you can deploy your Zappa application by executing:

    $ zappa deploy dev
```

Now that we have Zappa initialized, let's try deploying this to AWS. The truncated output is as follows:

```
zappa deploy
Calling deploy for stage dev..

Downloading and installing dependencies..

Packaging project as zip.

Uploading zappa-ipinfo-dev-1663345676.zip (6.0MiB)..

Waiting for lambda function [zappa-ipinfo-dev] to become active...
Deploying API Gateway..
Waiting for lambda function [zappa-ipinfo-dev] to be updated...

Deployment complete!: https://ohmkur5ipg.execute-api.ap-south-1.
amazonaws.com/dev
```

As you can see, the deployment returned an API Gateway URL. Let's hit it:

```
curl -s https://ohmkur5ipg.execute-api.ap-south-1.amazonaws.com/dev/
ipinfo/8.8.8.8|jq .country
"United States"
```

As you can see, this process was quite simple and Zappa took care of everything. I have never deployed a service this fast before! Let us see what happened behind the scenes after we entered zappa deploy:

1. Zappa packages the Flask app code and the virtual environment into a Lambda-compatible ZIP archive.

2. Dependencies are installed or modified to fit the AWS Lambda environment.

3. The handler is set up with a mechanism to translate between the API Gateway and WSGI.

4. The IAM policies and roles needed to create the resources are created.

5. The S3 bucket is created(if it does not exist) and the archive is uploaded to it.

6. A new Lambda function is created with the package in S3.

7. A new API gateway and routes are created that match the routes that Flask had defined.

8. The API Gateway is linked to the Lambda function.

9. The ZIP file is deleted from S3.

Now that we understand what Zappa can do, let's quickly look at a few more handy commands:

- To update the Zappa project, use `zappa update`.

- To see the deployment logs, use `zappa tail`.

- To see the status of the running Lambda/API Gateway resources, use `zappa status`.

- To roll back to the previous version, use `zappa rollback`.

- To remove all resources deployed, use `zappa undeploy` (ensure to run this at the end of the section).

Zappa offers a lot of additional features. The following are some of the most important ones:

- **Scheduling**: You can schedule a Lambda function to run on a specific schedule.

- **SSL support**: Deploy an SSL certificate from AWS/Let's Encrypt with custom domains.

- **Keep warm**: Use a schedule to invoke the Lambda function so that the container doesn't get cold.

- **Environment variables**: Pass environment variables to the function.

- **Static content**: You can use S3 to serve static content for your deployments.

- **API key/authorizer**: Support using the API key and Lambda authorizers with the API Gateway.

- **Events**: Allows your Python code to execute in response to events in SNS, SQS, S3, DynamoDB, and so on.

To learn more about Zappa, visit `https://github.com/zappa/Zappa`.

Zappa strikes a balance between the features it supports and the simplicity of using the framework. This is what makes Zappa unique and useful. In the next section, we will look into IaC using Terraform and how it can help with serverless deployment.

Infrastructure as code with Terraform

Cloud vendors have their infrastructure in data centers spread worldwide. When we create cloud infrastructure, we are interacting with the APIs these vendors have exposed to their customers. These APIs can be used to create and configure all services that the cloud vendors provide. As cloud infrastructure has become complicated, the work of managing it has also become complicated. This has opened up a new paradigm in infrastructure management – **infrastructure as code**, or **IaC**.

IaC is a new approach to infrastructure management and is a departure from the old ways of managing infrastructure manually and mostly through cloud consoles and CLIs. As the name suggests, IaC is all about maintaining your infrastructure using code. The desired end state of your infrastructure is written as code – this could be either a programming language or a **domain-specific language (DSL)**. Once the code is written in the IaC DSL, the IaC tooling will read through the code and implement the infrastructure in the cloud.

So, what are the advantages of using IaC? There are a few important ones listed as follows:

- Better productivity with automation
- Reduces human errors and outages due to misconfiguration
- Write once, deploy many
- Infrastructure management at scale
- Better change management life cycle
- Testing and policy enforcement with the CI/CD pipeline

When you switch to IaC, there are a number of improvements that you can make to the process of infrastructure management. Most of this can be applied by setting up a CI/CD pipeline for your IaC code. Now that we have a good idea of the benefits of IaC, let's start exploring the most popular solution in this category – Terraform. Terraform is an IaC solution created by HashiCorp and has become very popular among DevOps and cloud engineers. Let's start by understanding the foundational concepts and then try out an example.

Terraform concepts

The following are the key concepts of the Terraform DSL:

- **HashiCorp configuration language (HCL)**: HCL is the custom DSL used by Terraform.
- **Resources**: Resources are cloud resources that exist as independently configurable objects. These resources could be a Lambda function, an API resource, a VM instance, and so on. They are the building blocks of Terraform. It has its life cycle and attributes and can take input arguments and output attributes.
- **Providers**: Providers are Terraform plugins that interact with the APIs exposed by vendors. Each vendor will have a corresponding provider in Terraform.
- **Terraform configuration**: This is a file containing the definition of one or more cloud resources representing a piece of infrastructure using HCL.
- **Module**: This is a collection of Terraform configuration files that are reusable.
- **Data source**: This is a facility offered by each Terraform provider, to fetch external data.

- **State**: This is the snapshot of knowledge that Terraform has about existing cloud infrastructure.

- **Input variables and output values**: These are the parameters for Terraform configuration and the return values of a Terraform module.

- **Functions**: These are built-in functions provided by HCL.

There are other advanced concepts you will pick up when you dive deeper. The concepts we covered will get you started.

Terraform workflow

There are four core parts to a Terraform workflow. They are listed as follows:

- `Init`: This initializes the working directory with Terraform configuration for the first time.

- `Plan`: This builds a deployment/execution plan and allows you to inspect it. This could be considered as a dry run before actually applying the changes.

- `Apply`: This is where Terraform applies the execution plan to the cloud infrastructure.

- `Destroy`: This cleans up all the infrastructure that was created by Terraform previously.

Getting started with Terraform

First, you need to install Terraform on your developer machine. Follow the instructions at this link – `https://learn.hashicorp.com/tutorials/terraform/install-cli`.

In this exercise, we will reuse the same AWS account that we used in previous sections. Ensure you have an AWS account handy, with its CLI configured and the `AWS_ACCESS_KEY_ID` and `AWS_SECRET_ACCESS_KEY` environment variables, as follows:

```
export AWS_ACCESS_KEY_ID=<YOUR_ACCESS_KEY>
export AWS_SECRET_ACCESS_KEY=<YOUR_SECRET_KEY>
```

The structure of the Terraform file is as follows:

```
<block_type> [<parameter_1> <parameter_2> ...] {
<argument_1_name> = <argument value or expression>
....
}
```

A configuration file, that is, a file with the `.tf` extension, contains one or more Terraform blocks. Each of the blocks follows the syntax shown previously. The block has a type and zero or more parameters; the number of parameters depends on the type of the block. Possible block types are resource, variable, output, and so on. Blocks can have zero or more arguments, with values that are either plain values or an expression that evaluates a value. Blocks can also have nested blocks within them. We will learn more about this structure after we write our first Terraform file.

Before that, let us create a project directory:

```
mkdir tfproject
cd tfproject
```

Now, let us create our first Terraform configuration file and call it main.tf. We are going to create a configuration file that will create an S3 bucket and give it public/anonymous read access. The configuration file is as follows:

```
provider "aws" {
  region = "ap-south-1"
}

resource "aws_s3_bucket" "myweb" {
  bucket = "myweb-bucket-serverless101"
}

resource "aws_s3_bucket_acl" "website_acl" {
  bucket = aws_s3_bucket.myweb.id
  acl       = "public-read"
}
```

The file has three blocks, which are mostly self-explanatory, though we have briefly discussed them here:

- provider defines the cloud provider for which this file is written. It can also be used to input provider-specific settings, such as the AWS regions as you see in the file.

- The aws_s3_bucket resource defines an S3 bucket with the name myweb-bucket. It also has a myweb label/parameter that can be used to refer to this bucket elsewhere in the file.

- aws_s3_bucket_acl defines the ACL object. Its bucket argument points to the ID of the bucket we defined in the previous block and is labeled myweb.

Now, let us initialize the project:

```
~/tfproject$terraform init
Initializing the backend...
Initializing provider plugins...
Terraform has been successfully initialized!
```

This will create a .terraform directory in the current directory – which will store all Terraform-specific files and states. It will also create a lock file named .terraform.lock.hcl. The init function will also install the AWS provider package inside the .terraform directory as it is mentioned as the provider in the main.tf file.

Now that Terraform is initialized, let's validate the `main.tf` file:

```
~/tfproject$ terraform validate
Success! The configuration is valid.
```

Since the file is valid, let's proceed to do a dry run. The output has mostly been truncated as this creates a verbose output with every resource creation and every resource configuration for each of those resources:

```
~/tfproject$ terraform plan
Terraform used the selected providers to generate the following
execution plan. Resource actions are indicated with the following
symbols:
  + create
Terraform will perform the following actions:
  # aws_s3_bucket.myweb will be created
  + resource "aws_s3_bucket" "myweb" {
    + acceleration_status     = (known after apply)
.....TRUNCATED....
}
  # aws_s3_bucket_acl.website_acl will be created
  + resource "aws_s3_bucket_acl" "website_acl" {
...TRUNCATED...
Plan: 2 to add, 0 to change, 0 to destroy.
```

Now that we have seen the plan and it looks good, let's apply it. The `apply` command will first output the same output as the `plan` command and then prompt you to confirm it:

```
~/tfproject$ terraform apply
...TRUNCATED...
Do you want to perform these actions?
  Terraform will perform the actions described above.
  Only 'yes' will be accepted to approve.

  Enter a value: yes

aws_s3_bucket.myweb: Creating...
aws_s3_bucket.myweb: Creation complete after 2s [id=myweb-bucket-
serverless101]
aws_s3_bucket_acl.website_acl: Creating...
aws_s3_bucket_acl.website_acl: Creation complete after 0s [id=myweb-
bucket-serverless101,public-read]

Apply complete! Resources: 2 added, 0 changed, 0 destroyed.
```

So, Terraform tells us everything is successfully completed. Let's verify that using the AWS CLI:

```
~/tfproject$ aws s3 ls|grep myweb
2022-09-19 15:32:45 myweb-bucket-serverless101
```

As you can see, the bucket is created. Now, Terraform has stored the current state of your cloud as it knows it in a state file named `terraform.tfstate` in the current directory. If you want to see the current state, you can run the `terraform show` command. It is not safe to keep the state file in the local repo. Terraform has support for remote state files where the state will be written to a remote location, such as AWS S3 or Google Cloud Storage. There is also an option to carry out a remote state lock to avoid conflicts in modifying the state by multiple actors.

Now that we have seen how Terraform works, let's destroy the state so that it will remove the S3 bucket. Terraform will again present the current state with a destroyed status against the resources:

```
~/tfproject$ terraform destroy
...TRUNCATED...
Plan: 0 to add, 0 to change, 2 to destroy.
Do you really want to destroy all resources?

  Enter a value: yes

aws_s3_bucket_acl.website_acl: Destroying... [id=myweb-bucket-
serverless101,public-read]
aws_s3_bucket_acl.website_acl: Destruction complete after 0s
aws_s3_bucket.myweb: Destroying... [id=myweb-bucket-serverless101]
aws_s3_bucket.myweb: Destruction complete after 1s

Destroy complete! Resources: 2 destroyed.
```

This is just a taste of Terraform's capabilities. Terraform has a lot more constructs that will allow you to create complicated cloud infrastructure with ease. For more understanding, refer to the official documentation at `https://www.terraform.io/docs`.

If you are not using other serverless frameworks, Terraform is the best way to provision your serverless infrastructure using the same developer workflow and building pipelines as your code. In the next section, we will look into another IaC tool that is similar to Terraform but fundamentally different in a way.

Infrastructure as code with the Pulumi SDK

Pulumi has the same end goal as Terraform – use a developer workflow involving code to automate your infrastructure management. The fundamental way Pulumi is different from Terraform and many other IaC tools such as CloudFormation and ARM is the choice of language. Terraform and most other IaC tools use domain-specific languages such as HCL, YAML, and JSON, which allows you to declare the end state of your infrastructure, and then the IaC tool goes on to implement it. DSL is the preferred mode of achieving IaC because it is fundamentally a declaration of objects with several properties. The issue, however, is that, given it's a declarative model, there is not much custom logic you can employ. The best you can do is connect the resources using the constructs of your DSL or do rudimentary filtering and joining.

This is where Pulumi takes a different approach. Instead of depending on a chosen DSL, Pulumi supports IaC using popular programming languages. This allows you to automate your cloud infrastructure with the Pulumi SDK for your favorite language and chosen cloud vendor. Let us see what the advantages are of this approach:

- More control over the automation logic
- No need to learn a new DSL
- Improved developer productivity with a familiar language ecosystem
- Better testing with language native testing frameworks
- Built-in testing support

Let us get started with Pulumi and see for ourselves how different it is from Terraform. What we are going to do is use the Pulumi Python SDK for AWS and create the same S3 bucket.

Getting started with Pulumi

Install Pulumi by following the instructions at `https://www.pulumi.com/docs/get-started/aws/begin/`.

Once you have Pulumi and the Python runtime along with `pip` and `venv` installed, we can start. First, create your project directory:

```
mkdir pulumiproj
cd pulumiproj
```

Ensure that you have an AWS account and CLI set up before proceeding. When you are using Pulumi for the first time, if you run any commands, it will first ask you to log in to your Pulumi account to take advantage of their managed service. It is free for developers/individual accounts. If you want to avoid this, use `pulumi login -local`.

Now, let us create a Pulumi project. The `pulumi new` command is used for this and it takes a Pulumi template as an argument. A template is a combination of cloud and programming languages. In our case, this is named `aws-python`. Running the `pulumi new` command will prompt you to provide basic information about the project. All the questions are self-explanatory. For most of them, you can accept the default by pressing *Enter*:

```
~/pulumiproj$pulumi new aws-python
This command will walk you through creating a new Pulumi project.
project name: (pulumiproj)
project description: (A minimal AWS Python Pulumi program)
Created project 'pulumiproj'
stack name: (dev)
Created stack 'dev'
Enter your passphrase to protect config/secrets:
Re-enter your passphrase to confirm:
aws:region: The AWS region to deploy into: (us-east-1) ap-south-1
Saved config
Installing dependencies...
Creating virtual environment...
...TRUNCATED...
```

You can see that it asks for project and stack names. The project is the top-level organization of Pulumi code – like a Git repo. Stacks are like different containers/isolations of your infrastructure defined by a different configuration. This could be different deployment environments, such as dev/stage/prod or regional splits of your infrastructure.

The `pulumi new` command creates a directory scaffolding and dependency installations in the project directory. For Python, you will find a `venv` directory with all the Python dependencies installed. Other files are `__main__.py`, which is the main program that defines the infrastructure, and one or more stack -specific configuration files in YAML format. In this case, it will include `Pulumi.yaml` (for project-level config) and `Pulumi.dev.yaml` (for stack-specific config).

The `__main__.py` file has some sample code that creates an S3 bucket. It is as follows:

```
~/pulumiproj# cat __main__.py
import pulumi
from pulumi_aws import s3
bucket = s3.Bucket('my-bucket')
pulumi.export('bucket_name', bucket.id)
```

To learn how to configure public access to the bucket, you can refer to the API documentation at `https://www.pulumi.com/registry/packages/aws/api-docs/s3/bucket/`. We will modify the code and configure public access as follows:

```
import pulumi
import pulumi_aws as aws

bucket = aws.s3.Bucket("pulumi-test-serverless101",
    acl="public-read",
    tags={
    "Name": "Pulumi - Serverless 101",
    })
```

To learn more about the Pulumi AWS SDK, refer to the following link: `https://www.pulumi.com/registry/packages/aws/api-docs/`.

Now, test the code before actually deploying it:

```
~/pulumiproj$ pulumi preview
Previewing update (dev):
     Type                 Name                       Plan
 +   pulumi:pulumi:Stack  pulumiproj-dev             create
 +   └─ aws:s3:Bucket     pulumi-test-serverless101  create

Resources:
    + 2 to create
```

Now, that all looks good, so let's deploy the stack:

```
~/pulumiproj$ pulumi up
Previewing update (dev):
     Type                 Name                       Plan
 +   pulumi:pulumi:Stack  pulumiproj-dev             create
 +   └─ aws:s3:Bucket     pulumi-test-serverless101  create

Resources:
    + 2 to create

Do you want to perform this update? yes
Updating (dev):
     Type                 Name                       Status
 +   pulumi:pulumi:Stack  pulumiproj-dev             created
 +   └─ aws:s3:Bucket     pulumi-test-serverless101  created

Resources:
    + 2 created
Duration: 4s
```

Let us check whether the resources are created:

```
~/pulumiproj$aws s3 ls|grep serverless
2022-09-19 18:06:28 pulumi-test-serverless101-2a1b121
```

As you can see, a random string is added as a suffix to the name. This is due to the autonaming functionality of Pulumi. You can read more about it here – https://www.pulumi.com/docs/intro/concepts/resources/names/#autonaming.

You can remove this stack by running pulumi destroy.

As you can see, both Terraform and Pulumi made it easier to create cloud resources. You probably wouldn't notice a difference between the two in terms of the time and effort you put in. This is because we chose a straightforward use case for demonstration. As the infrastructure becomes more complicated and needs more customization, the differences will start to appear. The power of the SDK over DSL will be very obvious once you get past simple use cases. Also, remember that Pulumi also manages its state locally, but has options for remote state management like Terraform.

Now that was a simple introduction to Pulumi. There are a lot more features in Pulumi, but with the power of your chosen programming language, you don't need much more than this to start authoring your infrastructure. We covered different frameworks for serverless and IaC in previous sections. Now, let us move on to the next step – testing for serverless.

Testing serverless

Now, let us look at approaches to testing serverless. Similar to the rest of the sections in this chapter, we will be using AWS and Python as the references for the various topics we will cover here. Familiarity with testing, in general, is assumed. The following sections aim to provide guidelines and pointers rather than implementation examples.

Testing in serverless – challenges and approaches

Serverless boasts simplicity in development and architecture, but it poses a bunch of challenges while implementing testing strategies. A few major ones are listed as follows:

- Serverless has FaaS code that can be tested in the traditional ways of testing code, but the challenge arises because the code itself is tuned to be run in a specialized environment specific to the vendor/platform, and simulating those conditions becomes complicated when running locally or in a test environment.

- The application architecture is dependent on various cloud services and their behaviors. Emulating them in the test environment is very difficult.

- A lot of serverless architectures have some sort of asynchronous component, such as queues that deal with events. Testing the application flow in such scenarios is complicated.

- The interaction with cloud services makes integration tests more important and prevalent than unit tests, adding skewness to the testing pyramid approach. This leads to high maintenance costs for test setup and increases the overall testing time.

Now, let us look at specific testing strategies, such as unit tests and integration tests.

Local manual tests

While strictly not part of the testing pyramid, it is common for developers to test their code manually by running it locally and testing whether something breaks or works. In the case of Lambda, there are a few ways to do this:

- AWS **Serverless Application Model** (**SAM**) CLI: AWS SAM is a framework that AWS provides to develop serverless applications effortlessly. This is similar to the serverless framework, but only for AWS. When you are developing with SAM, it provides you with a CLI and you can invoke the Lambda function locally using the `sam local invoke` command.

- Similarly, the serverless framework supports local invocation using the `serverless invoke local` command.

- Zappa, while having an `invoke` option, can only be used to invoke the deployed function remotely. It is a similar case with `aws lambda invoke`.

Other serverless frameworks can do similar invocations for other vendors as well as for different languages. Depending on your development framework, you can try those out. Now, let us look into serverless.

Unit testing for serverless

Unit tests should be run on the developer machine or the CI/CD pipeline. They shouldn't have any interaction with external services. With FaaS code such as Lambda and other platforms, your entry point to the FaaS is always a handler function. All other functions written in the code are indirectly invoked by the handler or other functions. To better tune your code for testability, it is advised to separate the business logic into different functions and the interactions with external systems shouldn't be part of these functions.

It is easier to test the business logic functions if they are not dependent on external systems. You can pass test values or use mock objects to achieve this. Mock libraries and platforms provide you with libraries that can simulate the behavior of cloud APIs. For AWS, there are a few popular options. They are listed as follows:

- **Moto**: Moto is a Python library that can be used to mock the AWS Python SDK called boto. If your Lambda function is written in Python, you can include this in your Python test suite and mock the AWS API calls. If you are using another language, Moto allows you to spin up a local endpoint against which you can test your API calls. AWS SDKs provide an option to pass a custom API endpoint for all their functions. You can use this to redirect your API calls to the local Moto endpoint. For more information, refer to `https://github.com/spulec/moto`.

- **LocalStack**: LocalStack is a Python framework that allows you to spin up a local AWS API endpoint. Unlike Moto, it doesn't provide a mock library that you can include in the test suite. So, you have to pass the `localstack` endpoint (which you can run on your developer box or CI/D infrastructure) to the AWS SDK functions. LocalStack also provides a paid cloud service for more detailed testing.

- **Generic mocking of functions**: Test libraries in all languages have mock libraries that can be used to change the behavior of a function. For Python, there is a Python unit test library that also has a mock module within it. All mock libraries provide their own decorators that can be used to modify the behavior of a function you want to test.

- **Serverless offline**: The serverless framework provides a serverless offline plugin that emulates Lambda and API Gateway.

- **AWS SAM Local API**: The SAM CLI allows you to spin up an API endpoint locally.

- AWS provides local installations of some of its services, such as DynamoDB, which you can spin up to test locally. For S3, there is open source software such as **MinIO** and **Riak CS**, which have S3-compatible APIs.

There are more ways to do unit tests, but these should give you a good idea. Also, other cloud vendors have similar options, but not as many as AWS has.

Integration tests

Integration tests allow you to test your interactions with other components. With FaaS, these components are often other serverless services in the cloud. While we can mock the service call unit tests, integration tests should invoke their tests against the integration points (other serverless services or external systems). To do this right, the developer should set up a test environment within the cloud. This should cover all the cloud services that the FaaS code is interacting with. The typical workflow is the following:

1. Create the required cloud services in a cloud account – preferably in an account dedicated to testing/development.

2. To create these services, make use of an IaC framework. Define your requirement and configurations in the IaC code and launch the test stack in the cloud account.

3. Most of the serverless frameworks have the built-in capability to provision these services, either by using cloud APIs or the IaC code internally. If you are developing your FaaS with one of those frameworks, it will be much easier to set up the test environment.

4. Ensure that your code takes the endpoint details of all cloud services (such as S3 bucket names and DynamoDB table names) from environment variables. This way, it will be easy for you to configure your functions to run against any environment.

5. When writing test cases, avoid using mocks – which should be reserved for unit tests.

6. Business logic testing should also only be done in unit tests. Integration tests are supposed to flush out issues with integration points.

7. If your function talks to serverless data backends such as S3 or DynamoDB, ensure to load the required test data and clean it up after the test is done. Ideally, this should be part of your test fixture.

8. All function executions are triggered by events. When triggering the functions as part of the test, ensure you have stub events that you can pass to your functions. Depending on how the function has been triggered, the structure of the event will vary. Check out the corresponding cloud documentation to understand the event structure when writing your integration tests.

These are some of the broad guidelines for running integration tests in serverless applications. You should also follow up with acceptance/end-to-end testing once the function is deployed to your testing environment. While you might initially run unit and integration tests manually, this should ultimately be migrated to a CI/CD pipeline. We will take a quick look into CI/CD pipelines for serverless in the next section.

CI/CD pipelines for serverless

CI/CD for serverless is not very different from the normal CI/CD process. At the end of the day, you are going to build and test some code and then deploy it along with some services into the cloud. While there is complicated logic involved in this, it is good to ensure the entire process and pipeline are automated. Some of the common steps in the CI/CD stages are as follows:

1. Code commits with the right branching strategy.

2. **Linting**: Implement this for application code as well as the IaC code you might have as part of your project.

3. **Run security scanning for code**: Especially important in a cloud context, predominantly to prevent credentials exposure.

4. **Policy validations against IaC**: To see whether any of the service creations are violating organizational policies or resource limitations.

5. **Run unit and integration tests**: Locally run on the CI/CD machines, integration tests also need to deploy services to the test environment as a prerequisite.

6. **Build and package the application**: Usually, this is a self-contained ZIP archive of code and dependencies.

7. Deploy to production.

You can use the serverless frameworks and IaC tools for most parts of the CI/CD system. Where you run the CI/CD pipeline itself is a choice depending on your infrastructure. Given that cloud infrastructure can be provisioned and configured from anywhere, the choice is flexible.

Summary

In this chapter, we covered some of the serverless frameworks and IaC tools that can be used to make your serverless development and deployment easier. While writing short functions with minimal dependency is not a difficult task, using the right framework and CI/CD practices is essential when your use cases get more complicated. Most of the time, you can use your existing development process for serverless as well, as long as you understand the nuances of serverless infrastructure. In the next chapter, we will look into serverless security, observability, and best practices.

10

Serverless Security, Observability, and Best Practices

In the previous chapter, we saw how various serverless frameworks and automation work and how they deploy an application into production. The next logical step is to look into the post-production aspects of serverless computing, as well as some of the best practices in the serverless application life cycle.

Security is an important aspect of any running application. Software applications are susceptible to several security vulnerabilities. The more complex the application, the more attack surfaces it will have. This is why the cybersecurity landscape is so vast and diverse. In traditional application security, threats impact various layers of our application infrastructure, starting from the network, going through servers, and all the way up to the application. This is because a traditional application is hosted in a physical server within a data center premises. Since the entire infrastructure is owned by the application/business owner, the responsibility of securing all these layers also falls upon them. When using serverless computing, we pass on some of these security responsibilities to the cloud/ serverless vendor. Most of the time, we are left with the responsibility of securing FaaS and the partial responsibility of securing the serverless backends (as the core security of the BaaS service itself is with the cloud vendor). However, the bulk of the security vulnerabilities would remain at the FaaS code level, just like the application code vulnerability in a traditional environment.

While there are many vulnerability categories and their solutions are applicable to serverless (as we will see in the upcoming sections), it is also important to consider another aspect of running serverless applications – monitoring. Monitoring helps in troubleshooting, security auditing, and performance benchmarking serverless applications. While monitoring services for serverless computing is specific to each cloud vendor, we will cover some general guidelines and best practices, along with some examples from AWS.

In addition to security and monitoring, there are a few more best practices that will help us to run serverless applications more efficiently. We will cover some of these toward the end of this chapter.

In this chapter, we will cover the following:

- Security vulnerabilities and mitigation guidelines
- Serverless observability
- Serverless best practices

Let us start looking into the security aspects of serverless computing in the coming section.

Security vulnerabilities and mitigation guidelines

As mentioned earlier, serverless computing, by design, takes away a lot of security responsibilities from the application owner. A FaaS/BaaS vendor has to secure the underlying serverless infrastructure. However, this also opens up several other attack vectors for serverless applications. A bifurcation of security responsibilities between the vendor and the application owner is shown as follows:

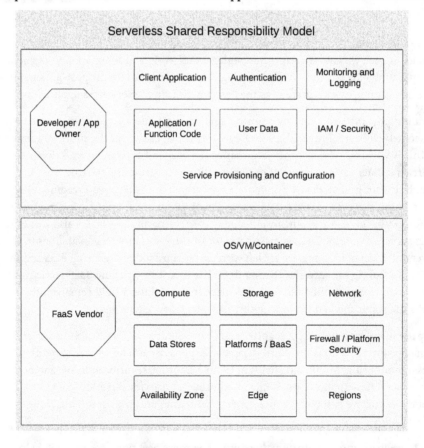

Figure 10.1 – Security responsibilities – the vendor versus the application owner

Serverless applications have several vulnerabilities, many of them overlapping with the vulnerabilities of traditional applications. The software industry and the cybersecurity community invest in identifying such vulnerabilities and coming up with strategies and techniques to defend against them. Often, such research is conducted and collated by various organizations and published for the community to adopt them. In this section, we will look at two such organizations and their recommendations on serverless applications – OWASP and CSA.

The OWASP Serverless top 10

The **Open Web Application Security Project (OWASP)** (`https://owasp.org/`) is a nonprofit organization that works toward improving software security, especially web applications. It is a community-driven organization that runs software projects and conducts educational and training programs. It helps developers to secure the web through community contributions. OWASP publishes many security guidelines and reference documents on various aspects of software security. One such reference publication is the **OWASP serverless top 10,** which details common serverless application security vulnerabilities. You can read this by visiting the following link: `https://raw.githubusercontent.com/OWASP/Serverless-Top-10-Project/master/OWASP-Top-10-Serverless-Interpretation-en.pdf`.

As the name indicates, this report lists 10 top vulnerabilities, listed as follows:

- Injection
- Broken authentication
- Sensitive data exposure
- **XML External Entity (XXE)**
- Broken access control
- Security misconfiguration
- **Cross-Site Scripting (XSS)**
- Insecure deserialization
- Using components with known vulnerabilities
- Insufficient logging and monitoring

OWASP constantly collects information from the community and the industry to keep their list up to date and reprioritized to reflect the current state of serverless security posture. These recommendations were first published in 2017 and, so far, remain unchanged. Before we go into the details of these vulnerabilities, let's look into the recommendations from another organization, called the CSA.

The CSA top 12 serverless vulnerabilities

The **Cloud Security Alliance (CSA)** (www.cloudsecurityalliance.org) is another nonprofit organization that drives security best practices in the cloud computing industry. It also provides education and training to developers and technologists to protect cloud resources. The CSA also operates with communities across the globe, led by a governing body of corporations, security practitioners, and other organizations. Similar to the OWASP top 10, the CSA published a list of the 12 most important vulnerabilities that can impact serverless applications. Let's look at them:

- Function event-data injection
- Broken authentication
- Insecure serverless deployment configuration
- Overprivileged function permissions and roles
- Inadequate function monitoring and logging
- Insecure third-party dependencies
- Insecure application secrets storage
- Denial of service and financial resource exhaustion
- Serverless business logic manipulation
- Improper exception handling and verbose error messages
- Legacy/unused functions and cloud resources
- Cross-execution data persistency

If you look at both these lists, you will find that most of the vulnerabilities in these lists overlap. So, instead of going through both lists, we will look at the common vulnerabilities one by one.

Event injection

Data injection in web applications happens when untrusted and malicious input is passed in the application logic directly. In traditional web applications, this usually happens via the web endpoint through direct user input. However, in the case of serverless applications, this is not always true. Direct input can be provided if your FaaS code powers an API Gateway endpoint, but most serverless applications take events from multiple sources other than API Gateway. Some of these events are as follows:

- Cloud storage events (uploading files to AWS S3, GCP Cloud Storage, and so on)
- Stream and message queue events (GCP Pub/Sub, AWS SQS, AWS Kinesis, and so on)
- Notification services such as AWS SES and SNS

- IoT events through IoT gateways
- Code commits
- NoSQL and SQL database events

As we saw in *Part 2* of this book, these event sources have different event schemas. This makes the input more complicated, with complex application code needed to support all types of events. There are different types of injections that are aimed at various components of the application stack. Some of them are listed as follows:

- Runtime code injection – depends on the language in which the FaaS code is written
- SQL/NoSQL injection – taking advantage of how database insertion queries work
- Object deserialization attacks
- **XXE** and **Server Side Request Forgery** (**SSRF**)
- Data tampering of messages in streams and queues

Some of the mitigation techniques for such attacks are as follows:

- Validating and sanitizing user input – never trust the data or assume its validity
- Restricting the privilege with which code runs – limiting its ability to harm the underlying runtime and OS
- Taking into account all possible types of event inputs that your FaaS code will handle
- For API Gateway/HTTP-based traffic, using a **Web Application Firewall** (**WAF**) to sanitize requests

Broken authentication

In sufficiently evolved serverless applications, several functions will serve different purposes and interconnect with synchronous or asynchronous workflows. Unlike traditional applications, there is no central control in this architecture. Each of these functions can be triggered by a variety of events from many sources. This creates a small but complex ecosystem, where implementing an authentication system that protects all involved resources is a hard undertaking.

Missing out on properly protecting a few resources can cause data breaches in a system very easily. This kind of vulnerability will result in exploitable resources, such as an API endpoint that is unprotected, or object storage buckets that leave files open to the world. Another attack surface is unprotected triggers that can launch functions. Consider a function that processes every incoming email in your support queue. An attacker carefully crafts a malicious email and sends it to the queue by spoofing the source address, resulting in the function getting executed with the malicious content.

This vulnerability is usually the result of the bad design of authentication schemes and access controls. It results in exploitable rogue entry points to a serverless application. This results in data leakage and breakage in business flow execution. Some of the guidelines to mitigate this are as follows:

- Do not reinvent the wheel and build authentication schemes.
- Use authentication services provided by the FaaS vendor or established third-party vendors. Some examples include the following:

  - Amazon Cognito
  - Azure App Service authentication and authorization
  - GCP Firebase authentication
  - Auth0
  - OneLogin
  - Okta

- Use secure API keys or SAML to protect API endpoints.
- For IoT workloads, use secure transport, one-time password, OAuth, and so on, depending on your platform and transport service.
- Use continuous verification of configurations and health checks to find open and vulnerable endpoints.
- Use third-party tools that enforce cloud security posture and compliance.

Insecure configurations

Insecure configurations are one of the top reasons why cloud resources in general and serverless computing in particular get exploited. Functions have associated timeouts for how long an invocation of a function can run. Similarly, there is also a concurrency limit that controls the number of function instances that can run in parallel. While this is quite useful in controlling our resource usage and performance, incorrect configuration of these settings – a low concurrency limit and long timeouts – can be exploited to run **denial of service** (**DoS**) attacks against your service. Another example of misconfiguration is cloud storage authorization and authentication. Such misconfigurations leave the cloud storage open, causing the leakage of sensitive data.

To mitigate this vulnerability, a serverless configuration scanning tool or cloud security posture management(CSPM) tool is recommended. There are also vendor-recommended security hardening configurations for storage and FaaS that should be adhered to. Data should be encrypted at rest and in transit. Using a cloud key management service to encrypt data and keep it rotating in defined intervals is a good practice.

Insecure secrets and credentials storage

Most serverless applications require you to access either data or other cloud services to deliver meaningful business value. This means that an application should have access to many application secrets, such as API keys and database credentials. A common mistake that happens in handling these secrets is that they are, oftentimes, stored as plain text in configuration files as part of your FaaS code. This means anyone with *read* access to the repo can obtain these credentials. It gets worse if the project is committed to a public GitHub repo, as it is a common practice for threat actors to scan public GitHub repositories for compromised credentials. Another comparatively less vulnerable way to store credentials is to keep them as environment variables. Although a much better approach than plain text storage, there are instances where these variables can leak through vulnerabilities in the FaaS code. The people tasked with deploying the application can also have access to these environment variables.

It is important to use centralized encryption key management for all your credentials. Cloud vendors offer their own solutions – for example, AWS Secrets Manager. For your encryption keys, you can also use AWS **Key Management Service (KMS)**. There are similar services provided by other cloud vendors. If you are on-premises in a data center, you can use a service such as HashiCorp Vault or a **hardware security module (HSM)**. If you have to keep your secrets in environment variables, ensure to keep them encrypted, and only decrypt them during runtime. Also, as a security practice, ensure to scan your Git repos regularly to see whether any secrets are committed along with code.

Broken access control

In the previous vulnerability, we looked at secrets to authenticate against other cloud services. Cloud security roles and policies define which function can have what kind of permissions on these services. Oftentimes, these permissions are *too permissive*, providing more privilege than needed for the function. Consider a function requiring *read* access to a cloud NoSQL database; it is quite common for developers to provide blanket *read* and *write* permission to the database. If the function with this privilege is compromised, the attacker can delete or manipulate the data, rather than only being able to read the data if restrictive permissions were applied.

This can often happen if there are common security policies applicable to different functions, where each of them might need a different set of privileges to access different resources. When the cumulative permissions for all these resources are applied to all functions through a common policy, a single vulnerable function can expose resources accessed by all services.

The way to mitigate this is to follow the principle of least privilege, allowing the minimum required privilege for each function. This helps in limiting the blast radius during a potential attack. Ensure that each function has its own dedicated role and privileges. You can also use security tools that will scan the security policies and flag potential overprivileged policies and roles.

Inadequate function monitoring and logging

A lack of monitoring and timely response allows attackers to achieve their objectives without being alerted about their intrusion. Most exploitations and attacks start with a reconnaissance phase, where the attackers slowly start probing an application for weaknesses until they find a vulnerability they can exploit. When there is sufficient monitoring and alerting that can provide early warnings about such reconnaissance missions, it will alert the application owners and engage their cyber defense team.

It is necessary to make use of the logging and monitoring capabilities that your cloud vendor provides you. However, you should also be mindful of the fact that these services, when enabled, have a limited default configuration to save cost and log the minimum information required for troubleshooting. However, a security audit requires much more verbose logging, with relevant details required to detect security breach patterns.

To mitigate this issue, proper monitoring and logging should be implemented for your functions. Add more details to your logging than provided by default by the vendor; you should add details of failed attempts to authenticate, insufficient permissions, and so on to your log. Use a logging library for your choice of FaaS runtime, and emit log-relevant log entries to standard output, which will be picked up by your cloud vendor's logging platform. Security teams use a centralized log analysis platform that is commonly referred to as a security information and event management (SIEM) system. You should channel logs from all your serverless functions and cloud service to the SIEM so that the security team can detect breach patterns in advance. We will learn more about monitoring functions in the upcoming sections.

Obsolete serverless resources

Over some time, all software systems accumulate legacy components, sometimes obsolete and abandoned. The same happens in serverless applications as well, as some of the functions become outdated or unnecessary. Similarly, the cloud resources used by these functions will also sometimes become obsolete, resulting in unused storage buckets and databases being abandoned. The issue with these obsolete resources is that, in addition to adding cost to an organization, it increases the attack surface of the serverless infrastructure. Obsolete resources have more chances of being exploited, since their security policy or code may not be up to date to prevent exploitation.

To mitigate this risk, organizations should periodically scan to detect unused resources and delete/decommission them. Using a **CSPM** solution or a **serverless security platform** (**SSP**) is recommended.

Insecure dependencies

Serverless functions by design have a small footprint in terms of resource usage, as well as the code used to write them. While these functions are aimed at performing one function, they can be dependent on third-party software and libraries. They could also be dependent on third-party web services. Irrespective of how secure and well written your FaaS code is, it can still be vulnerable if the third-party libraries are vulnerable.

To mitigate this risk, we need to have control of our dependencies. The following techniques will help with that:

- Scan your code, including the dependencies, for vulnerabilities.

- Remove unnecessary dependencies and upgrade the dependencies to stable versions.

- Use dependencies from trusted sources only.

- Automate all these processes and make them part of your CI/CD pipeline.

Improper exception handling and verbose error messages

Debugging serverless applications is a bit tricky, since you can't easily run functions locally with all the dependencies on cloud services. We have seen some of the ways we can test serverless functions in the previous chapter. But oftentimes, this doesn't suffice, and developers resort to adding verbose debug messages to the functions deployed to the cloud in production. When these verbose error messages are exposed to end users or threat actors who are running reconnaissance, it may reveal some of the inner workings or logic of the application, which can lead to exposing potential weaknesses or even data leaks.

To mitigate this, resort to using only standard debugging practices. Stick to the monitoring facilities provided by your vendor. On top of that, you can also use standard serverless frameworks that can extend such troubleshooting capabilities. The error message should not reveal any information about the internal implementations. If needed, add special error codes in the response body that only developers can identify and map to a particular type of internal issue.

Cross-execution data persistence

FaaS by definition has no persistent state across multiple runs of the same function. Vendors provide local state – in the form of storage, environment variables, and memory – for the duration of a single invocation of the function, as it will be needed for the interim processing of application data. This is fine, but an issue arises when vendors reuse the same container for subsequent invocations of the same function. This is usually done to avoid cold starts and improve performance. When the same container is reused, the state from the previous run might persist, and if the new run is to serve a different user or session, there is a possibility of corrupting or exposing the data of the previous run/user/session.

To mitigate this, all state information should be stored in remote cloud services – object stores, secret stores, databases, and so on. Developers should treat all data as sensitive and discard it at the end of the function run.

Insecure deserialization

Serialization is the process of converting objects and attributes in your application into flat format (a stream of bytes) that can be either saved into storage or transported between different application endpoints. Deserialization is the reverse process, where this stream of bytes is converted back to a clone of the original object. Deserialization can be exploited to inject malicious data into a running application. An attacker will either be able to tamper with the existing data by exploiting access control or use arbitrary remote code execution to modify the application logic and take over a service.

To mitigate this vulnerability, the only safe way is to accept objects only from trusted sources. If that is not possible, you should implement other safeguards, such as using digital signatures on serialized objects, strict type constraints, running deserialization code in low-privilege environments, and log serialization and deserialization for auditing.

Other common vulnerabilities – XXE and XSS

These are common web application vulnerabilities that apply to all web applications. XXE is a vulnerability that allows an attacker to compromise the application processing of XML data. Some mitigation methods use the standard XML processor from a cloud vendor's runtime/SDK, scanning code and dependencies for vulnerable packages, testing the API endpoints for XXE attacks, and so on.

XSS is an attack targeted at browsers that injects malicious scripts into trusted websites. There are a variety of ways in which an XSS attack can be carried out. There are several prevention techniques to mitigate these, covered on the OWASP website at `https://cheatsheetseries.owasp.org/cheatsheets/Cross_Site_Scripting_Prevention_Cheat_Sheet.html`.

Now that we have covered the security aspects of serverless computing, let's take a quick look at serverless monitoring.

Serverless observability

Monitoring and observability are two terms that are often used interchangeably. However, they are related and often complement each other. The Google SRE book defines monitoring as follows:

> *"Your monitoring system should address two questions: what's broken, and why? The 'what's broken' indicates the symptom; the 'why' indicates a (possibly intermediate) cause. 'What' versus 'why' is one of the most important distinctions in writing good monitoring with maximum signal and minimum noise."*

Monitoring allows you to understand the state of your applications and systems by using a predefined set of metrics and logs. The assumption is that we know what can potentially go wrong and how to look out for those problems. This essentially means you are watching out for a set of known failures. This has served us well in predictably managing our applications. The issue, however, is that as an application ecosystem gets complicated – with the number of distributed components in a microservice architecture – it gets challenging to predict the known failure patterns across all systems and the interactions between them. This is where observability comes into the picture.

Observability is theoretically defined as the ability to understand a system's internal state by analyzing the output it generates. These outputs include logs, metrics, and traces. Together, they are known as the three pillars of observability. As mentioned earlier, monitoring and observability complement each other. While monitoring answers the question of what is wrong with your software system, observability tells us why it is wrong. You can have monitoring without observability, although it provides only limited insights.

Now that we have a basic understanding of observability and monitoring, let's look into the challenges of serverless observability.

The challenges of serverless observability

Serverless computing by its very nature poses some fundamental challenges to observability. Functions are not always on and developers may have no access to the underlying infrastructure. In a traditional (non-serverless) computing model, we have access to the infrastructure, which allows us to deploy monitoring agents and daemons alongside a running application. Those agents would run in the background, without interfering with the application's request processing flow. They will keep collecting logs and metrics from the application and shipping them to remote collectors in their observability infrastructure.

Now, consider your FaaS code running in a serverless infrastructure. The only running process is your function running within a container. Anything and everything you want to run has to be implemented inside your function's handler. This means that there is no scope for a background agent to perform observability tasks. This leaves us with the observability tools and platforms that the cloud vendor provides us with. This will vary from vendor to vendor, but at a minimum, it usually includes logging the function invocations and the basic metrics of the function's performance and limits.

FaaS platforms manage the scaling and concurrency of our functions. One function invocation handles one incoming request and/or one unit of data processing usually. While this simplifies deployments and frees up developers from the hassle of handling concurrency, it also takes away the opportunity for developers to optimize code for concurrency. One such optimization from an observability standpoint is batching metrics and log processing. The observability daemons usually buffer the events and metrics they collect from multiple requests and collate them at defined intervals, before shipping them out. This greatly optimizes the performance of the observability stack. With functions, there is nothing to batch-optimize, and the data collection process has to be repeated for every invocation of the functions. This has grave implications in terms of performance and cost.

There is another fundamental problem here – the cost of observability. There is a cost to set up an observability infrastructure, but that is not what we mean here. We're referring to the penalty that users will have to pay in terms of the increased response time. Since data collection and shipping become an active part of your function invocation, it takes longer for the function to respond to a user request. This impacts the latency for a user-facing service, which in turn can have an impact on a business. Some daemons and frameworks that collect trace data have to first be bootstrapped as part of an application startup itself. This is required so that an agent can instrument the required classes in the startup, which can take a few seconds sometimes. In a traditional application, this was not a problem, since the app would start processing requests after the application was bootstrapped. The problem with functions is that for each incoming request, there is a new invocation of the function, and for each invocation, the agent bootstrap also needs to happen. This means that the bootstrap time is now added to the processing time of the request, thus increasing latency even before the data is shipped out.

Another common issue is that a lot of function invocations are through event-based mechanisms. Unlike an API Gateway-based function invocation that is synchronous, even processing triggered by queues and most other backends is asynchronous. While the invocations themselves can be tracked and observed, connecting them with their asynchronous triggers is a tough task.

As you can see, there are several challenges involved in serverless observability. Cloud vendors try to solve some of these challenges, although the maturity of the solutions varies from vendor to vendor. In the coming sections, we will see some of the observability solutions that major cloud vendors provide.

Serverless observability in AWS

There are multiple observability solutions provided natively by AWS. While these solutions are not usually specific to serverless computing, they are available to developers by default. In the coming section, let's explore some of these services and how they can help with observability.

Amazon CloudWatch

CloudWatch is the de facto monitoring platform for AWS. It has two components that deal with metrics and logs. Lambda by default is integrated with the CloudWatch platform. The Lambda platform collects several standard metrics and ships them to CloudWatch. Similarly, Lambda streams the details of each function invocation – this includes the invocation log as well as the output thrown by the function code itself. There are three minimum log statements per invocation (excluding your custom outputs) for the start and end of the invocation, as well as a summary report of the important resource usage. All log entries between the *start* and *end* log entries are the logs of a particular invocation.

Lambda reports many default metrics to CloudWatch. This allows you to set alarms on these metrics and keep an eye on your functions. The following are the metrics emitted by Lambda:

- The number of invocations
- The duration of invocations
- Error count
- Throttles to measure a concurrent limit breach

- Dead-letter errors to track a failure to write to dead-letter queues

- `IteratorAge` to measure the speed of event consumption

- `ConcurrentExecutions` for total concurrent executions

- `UnreservedConcurentExecution` for concurrent executions, excluding reserved concurrency

You can send custom metrics to CloudWatch from your code using the embedded metrics format of cloud-watching. This can be emitted as a log event, and CloudWatch will automatically pick it up and convert it into metrics.

AWS X-Ray

X-Ray is an AWS service that provides distributed tracing. It allows you to trace user requests through various components of your application. It does this by gathering data about these requests from each involved service and then stitching these data points together to form a trace. This trace can show the life cycle of a request as it travels through your application infrastructure. To learn more about X-Ray, visit `https://aws.amazon.com/xray/`.

Like other applications running on AWS, you can use X-Ray to trace Lambda functions as well. You need to enable tracing for each function that you want to trace. This can be done via the AWS Lambda console. Once enabled, for each invocation, Lambda will ring an X-Ray daemon that can receive the traces. In addition to the default details that the trace records, you can also use the X-Ray SDK to send custom trace data. Currently, the SDK is available for Go, Python, Java, Node, and .NET.

X-Ray also supports several AWS services, such as API Gateway, S3, and DynamoDB. The service that has integration with X-Ray will sample the request received by them and record the traces, which will be shipped to the X-Ray platform. A complete list of the services that support X-Ray is available at `https://docs.aws.amazon.com/xray/latest/devguide/xray-services.html`.

Serverless observability in GCP

Google Cloud has an observability platform called GCP operations suite. Formerly known as Stackdriver, it has the ability to integrate metrics, logs, and traces for applications and services running in Google Cloud. The operations suite has three component services:

- **Cloud Logging**: A real-time log management service that supports large-scale storage, analysis, and alerting.

- **Cloud Monitoring**: A fully managed platform to collect metrics and events from Google Cloud and several other vendors. It also supports collecting custom application metrics, liveliness checks, uptime probes, and so on. It also provides dashboards, charts, alerts, and so on for effective insights.

- **Cloud Trace**: Cloud Trace is a distributed tracing platform that collects latency data from applications. It will automatically analyze the collected data and generate detailed latency reports to help flush out performance bottlenecks.

Google Cloud also provides Cloud Profiler, which can profile the CPU and memory usage of your production applications. While not an observability tool, it allows developers to drill down and flush out performance issues of their production applications. GCP also has the Cloud Audit Logs platform, which is used to record administrative activities performed on the Google Cloud resources.

Google Cloud has two separate FaaS offerings – **Cloud Functions** and **Cloud Run**. Both are integrated with the operations suite. Cloud Functions by default offers simple runtime logging – these are the log entries emitted to `stdout` and `stderr`. For custom log entries beyond this capability, you should use the cloud logging libraries provided by GCP. You can also respond to specific log entries by forwarding these logs to another cloud function. Cloud Functions supports two types of audit logs – *admin activity* and *data access* audit logs.

Cloud Run also sends logs to Cloud Logging. There are two types of logs that are automatically sent from Cloud Run. One is *request* logs – these are logs of requests received by the Cloud Run services. Another one is container logs – these are emitted by the container instance of the Cloud Run service, mostly from your own application code running inside the container. Cloud Run also sends logs to audit logs. Both admin activity logs and data access logs are supported. For both Cloud Run and Cloud Functions, logs can be viewed from the cloud console, Cloud Logging Logs Explorer, the Cloud CLI, and programmatically.

Cloud Functions sends basic usage metrics to cloud monitoring. These include execution counts, execution time, and memory usage. API calls, API call quota, and function execution quota are also monitored. You can set alerts on breaches of these metrics. Cloud Run is also integrated with cloud monitoring and provides performance monitoring, metrics, and uptime checks. You can also create custom metrics for Cloud Run using a log-based metrics approach. The list of metrics supported by Cloud Run is found at `https://cloud.google.com/run/docs/monitoring`. You can view your metrics using Metrics Explorer, the cloud console, the CLI, or programmatically.

Distributed tracing with Cloud Trace is supported only for Cloud Run services. The incoming requests to these services are automatically traced. This allows you to identify latency issues in your request path without additional instrumentation. You can also add custom instrumentation that will allow you to use Cloud Trace to instrument the time taken by your application to talk to different downstream dependencies. Not all requests will be traced by Cloud Run – it is limited by the sampling rate enforced by GCP, which is 0.1 requests per second for each container instance.

In addition to these services, there is also another product called Error Reporting. This service aggregates runtime errors caused by Cloud Run or Cloud Functions and reports them.

Serverless observability in Azure

Azure Monitor is the observability platform of the Azure cloud. It stores observability data in three different data stores – metrics, logs, and changes. The log store also stores the trace data. Azure Monitor provides aggregation, analysis, and visualization of these data types along with response capabilities. Azure Application Insights is an extension of Azure Monitor that provides an **application performance monitoring (APM)** service. It allows you to understand how an application is performing.

Azure Functions is integrated with Application Insights by default. It collects logs, performance, and error data from the function executions. In the selected language runtime, you can also use the Application Insights SDK to write custom telemetry data. For logging, you can stream execution logs using the built-in log streaming or the live metrics stream. Similar to Azure Functions, we also have Azure Container Apps in the Azure serverless land. Consider it the equivalent of Google Cloud Functions and Cloud Run. Like Functions, it also integrates with Azure Monitor for metrics, alerts, and log analytics.

This concludes our section on serverless observability. Other than the vendor-specific observability tools that we covered here, there are several observability vendors that support serverless platforms. The most notable ones are as follows:

- New Relic
- Splunk
- Dashbird
- Lumigo
- Epsagon
- Thundra

There are many more vendors other than these. Checking out some of these vendors will give you an idea of the capabilities the platforms offer. In the next section, let's look into serverless best practices.

Serverless best practices

While all FaaS platforms offer fundamentally similar services, the way you can configure and fine-tune them differs from platform to platform. For this reason, there are several platforms and framework-specific best practices. We will cover a list of best practices next; some of them are common to all vendors, while others are specific to some vendors and/or serverless features:

- **One function per task/route/event**: FaaS functions should be simple and performant to get the most out of the serverless paradigm. To do that, ensure that functions serve only one use case. Restricting the scope of a function also helps in easier debugging and scaling. If you are using event processing architecture, ensure that your function only processes one type of event, rather than an array of event types. If your function serves an API Gateway endpoint, limit it to process one route/REST resource along with its verbs. For all other use cases, limit the function scope to one task that serves the smallest business use case. Also, unless necessary, don't have your function call another function.

- **Use IaC for provisioning and configuration management**: Your functions involve both application code and infrastructure. While code constantly changes, your infrastructure components and configurations ideally should be determined and locked down, with limited changes, and only when necessary. When code has to be changed, you should have the ability to track the changes. This can be achieved by adopting infrastructure as code for your infrastructure and configuration needs.

- **Load-test your functions**: Functions are usually billed for the resources they use per invocation. To ensure your functions are optimized for cost, it is necessary to load-test your functions and determine the baseline performance. This will help you to set up realistic memory and CPU limits for the functions, leading to cost optimization.

- **Configure DLQ for event-processing functions**: When processing events using functions, there is a chance that some of the events might cause a function to fail. This could be due to a wide variety of reasons, including wrongly formatted messages or oversized messages. In such cases, debugging or reprocessing an event will be necessary, and for this, the failed events need to be stored away. This is where the **dead-letter queue** (**DLQ**) comes into the picture. DLQs are the destination for all such failed messages. This usually comes as a feature of the message queue service itself. For example, in Amazon SQS, when you provision a message queue, you can also add a DLQ to it. Some of the other services that offer DLQs are Amazon EventBridge, Apache ActiveMQ, Google Pub/Sub, Azure Event Grid and Service Bus, RabbitMQ, Apache Kafka, and Apache Pulsar.

- **Avoid wildcard privileges in IAM policies**: It is quite common to assign high-privilege security policies to serverless functions during the initial development process. This is usually due to developers still discovering the service dependencies of an application and the privileges it requires. The IAM policies in such cases contain wildcard access to certain resources or resource types. For example, consider a function that requires read permission to an AWS DynamoDB table. The developer might create an IAM policy that allows full access to all DynamoDB tables in the account or, if we are lucky, all privileges to a specific table. What usually ends up happening is that the same privileges are carried over to production. If the function is compromised somehow, this will allow the attacker to gain much higher privileges and inflict more damage. To decrease the blast radius, the IAM policies should never contain wildcard privileges. Rather, the policy should specify the particular cloud resource with the most granular and specific privilege that it requires.

- **One IAM role per function**: Serverless frameworks or cloud automation tools will create common IAM roles and associated security policies for all the functions they create. While this works well at the beginning as the number of functions increases and their use cases vary, this will get out of hand. The common policy and the role will grow to accommodate more use cases, accumulating more privileges to several resources that only a handful of functions would need. Therefore, when going into production, it is strongly advised to use a dedicated IAM role and policy for each function.

- **Store secrets and credentials in external secret stores**: Most functions don't work in isolation; they need access to external services to either fetch data or post data. This requires that a function should have the credentials necessary to authenticate with these external services. It is not advisable to store these as part of your application deployment artifacts, code, or configuration. You should store these credentials in a secret store that is either provided by the function vendor or a third party. For example, AWS provides Secrets Manager, allowing developers to securely store the credentials in this secret store in an encrypted format. Access to the secrets of Secrets Manager is controlled by IAM policies and roles. Your function should have the right role and associated policy that allows permission to access these secrets. It is also recommended to rotate these secrets at defined intervals.

- **Import only the minimum required packages**: In most cases, your function code will depend on a lot of libraries. There is a common practice to import large libraries fully into code, but the code will only use a few methods or classes from these libraries. Importing large libraries has an impact on startup time as well as the memory footprint. In a FaaS environment, both are critical resources and should be kept to the minimum number required. So, for all required dependencies, ensure to import only what you need.

- **Code security scanning**: It is ideal to scan your code for common vulnerabilities; these could be in included libraries or your application code. It is also a common pitfall to embed your credentials in code. There are open source and vendor-provided security-scanning solutions that can detect this for you. Ensure to integrate this in your build process.

- **Enforce strict timeout**: Most vendors allow you to set the timeout within which your function should finish execution. it is quite common for developers to set the timeout to the maximum allowed time, due to a lack of understanding of the function's evolving needs. It is a good idea to find out the right timeout that works for your functions and enforce it. It helps in two ways, the first being cost control with a shorter runtime. The other advantage is that if your function gets compromised, attackers will have limited time to run their exploit compared to the large default timeout.

- **Shift left – use a CI/CD pipeline**: While functions themselves are simple and ideally contain a small number of lines of code, managing the developer workflow of continuous updates and builds can become tiring and error-prone. It is recommended to bring in a CI/CD system to automate this task. It will also allow you to implement other best practices, such as automated testing, code scanning, and policy enforcement, without hassle.

- **Use a WAF**: If your functions are exposed via an API Gateway endpoint, there is a good chance that attackers will discover this eventually, especially if your application/business is popular. The attackers might try to launch a **distributed denial of service (DDoS)** attack against your endpoint. This will cause your service to go down due to resource exhaustion, or costs to go through the roof due to autoscaling. Protecting your high-volume web endpoint with a WAF is a good way to protect against DDoS as well as other web application exploits.

Summary

In this chapter, we learned about the security best practices that will protect your serverless applications. We also looked at the services that cloud vendors offer for the observability of your serverless applications. Finally, we looked at the most common best practices that will help you to design and develop your serverless web applications better.

In the next chapter, we will look into some of the common architecture patterns and serverless solutions.

11

Architectural and Design Patterns for Serverless

In this section of the book, so far we have looked at software frameworks, infrastructure as code, security best practices, and observability. All these provide a good foundation for developing serverless applications. In this chapter, we will look at another aspect of serverless development that aids in the design of serverless applications – design patterns.

Design patterns are generic, repeatable solutions to common problems. These are not solutions that can be readily converted to code, rather they provide a high-level framework that can be adapted to the problem at hand. Consider them as templates that provide a combination of best practices and reference architecture. In software engineering, there are several foundational design patterns that were proposed and well established decades ago. These design patterns apply to all software engineering problems and are usually associated with object-oriented design.

A design pattern provides a reference template to solve problems and enables software engineering teams to speak a common language while attempting to solve real-world problems. This is why patterns are even more important. It allows different and disconnected software teams solving the same problems to use proven solution frameworks that came from the collective wisdom of the software community. While the original design patterns were based on object-oriented software design, with the advent of the cloud, a new set of patterns emerged – **Cloud Design Patterns**. These patterns are focused on providing a framework to build reliable, scalable, and secure applications in the cloud. Cloud design patterns have been evolving as more and more services and vendors flock into the cloud domain. As serverless became mainstream and matured to handle production workloads and business problems, serverless design patterns became an integral part of cloud design patterns.

We will cover the following topics in this chapter:

- Design patterns primer
- Architectural patterns
- Cloud architecture patterns – vendor frameworks and best practices

- Three-tier web architecture with AWS

- Event-driven architecture with Azure

- Business process management with GCP

- More serverless designs

- Serverless applications in the Well-Architected Framework

Design patterns primer

The concepts of design patterns first appeared in the book *A Pattern Language: Towns, Buildings, Construction* (Christopher Alexander,1977). Influenced by this, in 1987, Kent Beck and Ward Cunningham authored a paper on using pattern languages for object-oriented design. While this kicked off the discussion on design patterns, it became mainstream with the publication of the book *Design Patterns: Elements of Reusable Object-Oriented Software* (Erich Gamma, Richard Helm, Ralph Johnson, and John Vlissides, 1995 – Addison-Wesley Professional). The authors are popularly referred to as The Gang of Four and the book is called the *Gang of Four* book.

The *Gang of Four* book divided design patterns into three categories. The categorizations were based on the life cycle and interaction of objects. They are **creational**, **structural**, and **behavioral** design patterns. There are a total of 23 design patterns under these categories. The following tables show the design patterns in each category:

Creational	Structural	Behavioral
Singleton	Adapter	Chain of responsibility
Builder	Bridge	Command
Prototype	Composite	Iterator
Factory Method	Decorator	Mediator
Abstract Factory	Façade	Memento
	Flyweight	Observer
	Proxy	State
		Strategy
		Template Method
		Visitor

Table 11.1 – Categories of design patterns

Now let us look at the first category of design patterns – creational design patterns.

Creational design patterns

Creational design patterns deal with class instantiation and object creation. It helps in reducing complexities arising from object creation by implementing controls around how object creation is done. There are two classifications within this pattern, **object creational patterns**, which handle object creation, and **class creational patterns**, which deal with class instantiation. The following five are creational design patterns:

- **Singleton pattern**: One of the simplest design patterns, singleton ensures that only one instance can be initiated from a given class. When an object creation request is invoked, it either creates the object (for the first time) or returns the existing object.

- **Builder pattern**: This pattern separates the instantiation of a complex object from its representation. In such a case, the instantiation itself will be handled by another object called the builder.

- **Prototype pattern**: This pattern is used when the type of object to be created is modeled by a prototype instance, which can then be cloned to create new objects.

- **Factory pattern**: This method is used to create an object without specifying the exact class of the object that will be created. A factory method is implemented as an interface and implemented by the child classes. Alternatively, it is implemented in the base class and possibly overridden in derived classes if necessary.

- **Abstract factory pattern**: Unlike factory patterns, the abstract factory pattern is used to create families of related or dependent objects. It encapsulates a collection of factories with a common theme and doesn't specify the individual classes.

Structural design patterns

Structural design patterns help in establishing relationships between objects and entities. The idea is to use these relationships to integrate into larger composite structures. The following are the seven structural design patterns:

- **Adapter pattern**: The adapter pattern creates an adaptor that makes two incompatible entities work together. This enhances the reusability of the code.

- **Bridge pattern**: The bridge pattern is useful in separating the abstraction from its implementation so that both can be changed independently. It allows changes to be made in the implementation without affecting the clients.

- **Composite pattern**: The fundamental idea of the composite pattern is that a group of objects can be composed into tree structures so that they can represent part-whole hierarchies. This is useful when your application deals with hierarchical data structures that share common denominator functionality that works across the structures.

- **Decorator pattern**: The decorator pattern provides flexibility to add additional capabilities to an object dynamically. This is done without affecting the behavior of other objects created from the same class. The decorator classes provide additional functionality by wrapping the original class in it.

- **Facade pattern**: The façade pattern hides the complexity of the other interfaces and classes by providing a simplified interface. This essentially decouples the client from the underlying complexities.

- **Flyweight pattern**: This pattern recommends the reuse of existing similar kinds of objects by storing them. This reduces the number of objects and helps improve the performance with lower RAM and storage usage.

- **Proxy pattern**: This pattern provides a substitute object to cordon off access to the original object. It is helpful in controlling access to the original object and providing only the features that we consider necessary to be exposed. There are several implementations of proxy patterns, *remote proxy* and *virtual proxy* being the most common ones.

Structural patterns are ideal when the existing system is too complex to be refactored, but the user needs it to be simplified. This is done by analyzing the complexity and relationships between objects and assembling them together to provide a much more streamlined view.

Behavioral design patterns

Behavioral design patterns deal with the interactions and responsibilities of objects, in other words, the behavior of the objects. The fundamental idea is that the interactions between the objects should be simplified while avoiding tight coupling between them. The following are the ten behavioral design patterns:

- **Chain of responsibility**: The chain of responsibility or chain of command pattern avoids the coupling of the sender of a request to its receiver. This is achieved by giving more than one object to handle the request.

- **Command pattern**: This pattern decouples the object that invokes an operation/command from the object that knows how to perform the operation.

- **Iterator**: The iterator pattern allows you to process items in a collection without revealing the representation of the data underneath. It provides access to the items without disclosing the inner structure of the data.

- **Mediator pattern**: This pattern acts as a bridge between objects by detaching them with a mediator object. This helps in avoiding components being strongly coupled with each other. This brings down the complexity of communication between a large number of objects.

- **Memento**: The memento pattern is needed when an action needs to be performed on an object but with the possibility of being undone if necessary. This requires that there be a way to save the state of the object being modified and later restored.

- **Observer**: This pattern observes the objects in one-to-many relationships and when the state of an object changes, all the dependent objects are informed and updated.

- **State pattern**: The state pattern allows an object to modify its behavior when its internal state changes. This works similarly to finite state machines.

- **Strategy**: This pattern can be used to select an algorithm at runtime. The algorithms will be interchangeable and independent of the client that uses them.

- **Template method**: This is a method in a superclass that implements an algorithm with a skeleton of high-level steps. The individual steps are implemented by additional helper methods implemented either in the same class or subclasses.

- **Visitor pattern**: This pattern allows defining a new operation without modifying the existing object structure. The logic of these operations is moved to a different class called a visitor.

That concludes our list of design patterns. This introduction was provided to give you an idea of how repeatable best practices can help teams design better software and speak a common design language. But design patterns are not without their problems and have attracted criticism. Sometimes people can be lost in implementing the pattern verbatim – they lose focus of the problem that they want to solve and compromise on efficiency. Another common pitfall is the overuse of patterns, where they are not necessary and where simpler methods would have sufficed.

While these patterns are the foundation of all designs from the object-oriented design perspective, many other patterns have come up as design practices evolved. There are architectural patterns and integration patterns. We will look at them briefly in the following sections.

Architectural patterns

Software architecture is a way to define structures and systems that describe a software system. Wikipedia defines **architectural patterns** as follows:

> *An architectural pattern is a general, reusable solution to a commonly occurring problem in software architecture within a given context.[1] The architectural patterns address various issues in software engineering, such as computer hardware performance limitations, high availability, and minimization of business risk. Some architectural patterns have been implemented within software frameworks.*

As you can see, this definition is not very different from the one for design patterns, except that it applies to software architecture as opposed to OOPs. A pattern is not an architectural implementation, but rather a concept that defines basic elements to solve a category of problem. A pattern can be implemented in different architectures. The following is a list of the most common architectural patterns that are used today along with their standard definitions:

- **Broker pattern**: This pattern is used to decouple components of distributed systems that talk to each other using **remote procedure calls** (**RPCs**).

- **Event-Driven Architecture (EDA)**: EDA, as its name suggests, deals with the discovery of, consumption of, and response to events.

- **Microservices**: Microservices decouple a complex application into loosely connected single-purpose services that communicate through lightweight protocols.

- **Service-Oriented Architecture (SOA)**: A predecessor to microservice architecture, SOA also breaks down services into smaller discrete services, but the complexity of these services can range from simple to complex, and they often use the same protocol and use messaging middleware.

- **Model-View-Controller (MVC)**: The MVC pattern separates the internal representation of information from the user interface and how the information is exposed.

- **Multitier architecture** (**3-tier, n-tier**): This pattern bifurcates application processing logic, data management, and presentation so that they can be implemented as separate software stacks. The most common implementation of this architecture is 3-tier architecture.

- **Peer-to-peer**: This pattern comprises equally privileged participants (usually application servers) that distribute tasks between themselves.

- **Publish-subscribe pattern**: In this pattern, messages that are relevant to applications are produced by a party called a publisher without the prior knowledge of who will consume them. The subscribers are applications that express interest in consuming a specific class of messages and do so through message subscriptions.

This is only a limited subset of all architectural patterns. There are other patterns, such as **enterprise application integration** patterns, which improve connectivity between enterprise applications. There are three categories of enterprise application integration: **access**, **lifetime**, and **integration**. Getting into these aspects of software design merits its own book. This information is provided so that new developers will get a basic understanding of the purpose and history of patterns.

Now that we have a foundation of design and architectural patterns, we will get to the core focus of this chapter – cloud architectural patterns and, by extension, serverless patterns.

Cloud architecture patterns – vendor frameworks and best practices

All cloud providers have been constantly churning out services and improving existing services to meet increasing customer needs. Just like most programming languages, you can use different approaches and different components to do the same things. The variety of choices and feature explosion is making it hard for customers to choose the optimal solution. This can result in sub-optimal cloud solutions and has an impact on the business value customers can derive from their cloud adoption.

Amid these overwhelming options and confusion, the top three cloud vendors – AWS, GCP, and Azure – have produced cloud architecture frameworks to aid their customers. The aim is to lay out best practices and architectural solutions that address common business problems. The cloud vendors named their frameworks as follows:

- **Amazon Web Services**: AWS Well-Architected Framework (`https://aws.amazon.com/architecture/well-architected`)

- **Google Cloud**: Google Cloud Architecture Framework (`https://cloud.google.com/architecture/framework`)

- **Microsoft Azure**: Azure Well-Architected Framework (`https://learn.microsoft.com/en-gb/azure/architecture/framework/`)

While these frameworks came out at different times, (with AWS releasing the Well-Architected Framework in 2012), they undergo regular revisions and additions to keep up to date with their changing cloud portfolio and customer needs. AWS is already on its ninth revision! Each framework is organized into a number of pillars. The following table summarizes the pillars of all three frameworks:

AWS Well-Architected Framework	GCP Google Cloud Architecture Framework	Azure Well-Architected Framework
Operational excellence	Operational excellence	Operational excellence
Security	Security, privacy, and compliance	Security
Reliability	Reliability	Reliability
Performance	Performance optimization	Performance efficiency
Cost	Cost optimization	Cost management
Sustainability		

Table 11.2 – Cloud architecture framework pillars comparison

These pillars apply to cloud architecture and as a subset, to serverless architecture as well.

These are guidelines for how to approach a problem, and the documents at the preceding links will give you a good idea of how to design for the cloud. What we are more interested in in this chapter is looking at solutions for common problems using cloud services (with a focus on serverless services). In the following section, we will look into some of the **cloud architecture patterns** or solutions. Some of them are pure serverless solutions while some are non-serverless or semi-serverless solutions. A walk-through of these varied solutions should give you a good idea of how to engineer for the cloud and where to fit in serverless solutions.

A discussion on cloud design patterns can be had in two ways: one is to use abstract or generic concepts and provide a framework that you can apply to any cloud; the other way is to present a solution using one of the cloud vendor solutions. Given that we have covered multiple clouds in the chapters in *Part 2*, the expectation is that the second approach will connect well with readers. Hence we will follow that path.

Three-tier web architecture with AWS

Three-tier architecture is one of the most popular architecture patterns for web applications. It divides the entire stack into three:

- **Presentation tier**: This is the user interface that your customer interacts with. In a traditional web application, this will be frontend built-in web technologies such as *HTML, JS, CSS*, and so on.

- **Application tier**: Also known as the logic tier, this handles the business logic of the application. It processes information from the presentation layer or returns information to it. It also works on data to compute its business logic. This is what you typically call a backend service and is usually written in services such as *Java, Go, Python*, and so on.

- **Data tier**: Sometimes called the database tier, this tier stores and serves all the data that the application needs. This backend generally is SQL or NoSQL services such as *MySQL, MongoDB*, and so on.

The following diagram depicts the concept of three-tier web architecture:

Figure 11.1 – Three-tier web architecture

Let us first look at how you can use AWS non-serverless services to set up such a web architecture:

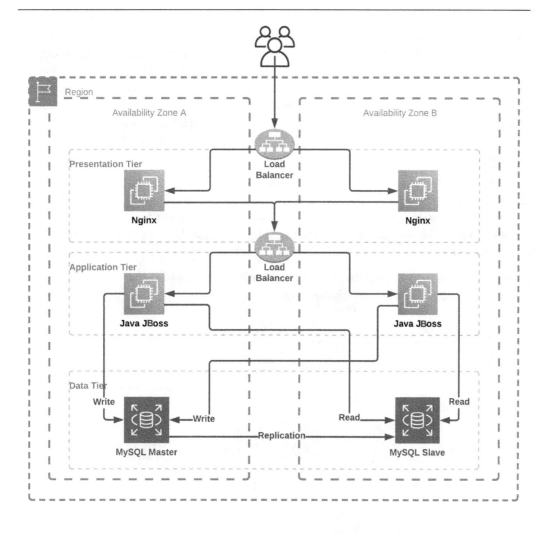

Figure 11.2 – AWS three-tier web architecture with VMs

The entry point to the application is an external load balancer, behind an array of nginx servers running on EC2 instances exposing a static web application. This is what is presented to the user and hence it is named the **presentation layer**. Any requests that are needed to run business logic are passed on to the **application layer**, which has an internal load balancer as its entry point, backed by an array of JBoss servers running a Java application. Like nginx, these apps also run on a set of EC2 instances. Next comes the **data layer** where all data is stored in a MySQL cluster with master and slave servers. All writes are taken by the master while the slave serves the read requests. All writes are replicated from the master to the slave.

Now, this is a simplified architecture and fulfills its purpose very well. But it doesn't exactly fully represent modern-day cloud architecture. The architecture can be enhanced with a number of cloud services. In the data layer, we could use an AWS RDS-managed *MySQL* service. We could also throw in a managed `Redis` cluster if the caching of data would help the read path. In the presentation layer, we can add the CloudFront CDN service in front of the load balancer to cache static web assets, along with the **Web Application Firewall (WAF)** solution if you want protection against DDoS and other attacks. If your customer base is across the globe, you might want to replicate the same setting in multiple regions – which will come with its own challenges.

The extent of additional services and enhancement you add to this architecture will depend on factors such as the size and geographical distribution of your user base, the amount of data you handle, and so on. But most importantly, it will also depend on your budget. That is why there is no one-size-fits-all solution here and cloud architects and developers have to make the right choices to meet their business demands.

Now that we understand this architecture, let us look at how this can be translated to a serverless solution using AWS services. See the following figure:

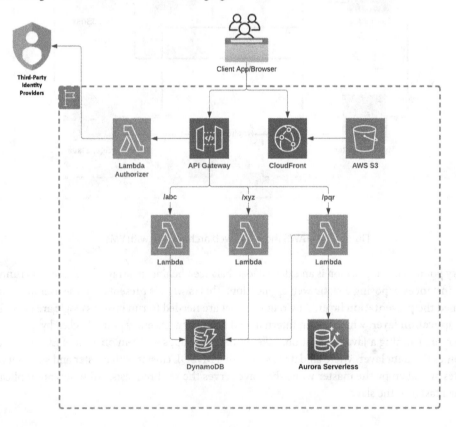

Figure 11.3 – AWS serverless three-tier web architecture

The presentation layer is replaced by an S3 bucket with all the web assets that can present a web UI. This is cached and distributed for speed and efficiency using the CloudFront CDN. Web assets from this presentation tier work out of the users' web browser, with JavaScript providing dynamicity and necessary communication to the application layer.

The application layer is an API gateway. It exposes one or more APIs that will serve various independent units of business logic that are handled by different Lambda functions. A custom Lambda authorizer is also used for integrating external authentication.

The data layer in the VM based architecture diagram was MySQL/RDBMS. While AWS provides a serverless version of its Aurora DB offering, it may not always suit our needs. In that case, you might either want to use RDS (non-serverless) or your own MySQL cluster. In this case, the Lambda function has to be deployed with access to your VPC (aka production network) so that it can access the MySQL servers running in your network. It is advisable to move your data from SQL to NoSQL databases, which are easier to manage and offer as a serverless variant. It may not be possible to move all the data from SQL to NoSQL, so migrate all data that doesn't require transactional guarantees that RDBMS offers. For the rest of the data, either use Aurora Serverless on RDS or a self-hosted DB in your VPC.

Like the original non-serverless design, a number of enhancements are possible here. For authentication, you can use **AWS Cognito**, which can replace your custom Lambda authorizer Lambda function and do much more. You can introduce **AWS ElasticCache** to query any data that you might have. You can use Lambda at the edge to run certain data transformations before it even hits your API gateway. You might need to integrate with **AWS Route 53** for DNS resolution and a certificate manager for handling SSL certificates.

This should have given you an idea of how to translate a non-serverless architecture solution to a serverless-oriented one. While this is an exciting possibility, be aware that a full-fledged serverless solution may not always be necessary, and mixing and matching the right services to reduce cost and improve reliability by following the Well-Architected Framework is the way to go. We will look at another popular architectural pattern next – event-driven architecture.

Event-driven architecture with Azure

An **event** represents a change in the state of or an update to a system or a unit of data that is typically triggered by a user action. Events contain the state of an item or identifiers to look up the information that leads to the state change or update. **Event-Driven Architecture** (**EDA**) helps with decoupling different but related business processes, thus facilitating the scaling of each process independently. The central theme of this architecture is an event router that receives events from microservices (event producers) and delivers them to event processing microservices (consumers). The idea is shown in the following figure:

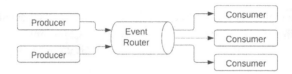

Figure 11.4 – Event-driven architecture

To understand this better, let us consider a simplified e-commerce transaction where you purchase an item from an e-commerce application. The moment you make payment, the purchased item is reserved for you, and your order is confirmed. This order confirmation is an event that will contain the item that you purchased and the address to which it is to be delivered. This event will be processed by your order microservice, which will in turn create another event with the item details and the warehouse that should process it. The warehouse microservice receives this event, creates an instruction to package the item, and deducts the item from the warehouse inventory. It then creates another event for the shipping partner to pick up and deliver the item. The delivery microservice will receive this event and notify the shipping partner to pick up and deliver the item. The shipping microservice will keep track of the item until its delivery. Once the item is delivered, the shipping microservice will mark the order as completed. All the while, there will be a notification microservice that also processes these events and keeps updating the customer.

While all these microservices work in tandem to deliver the order to you, none of them waits for the other service to complete its part of the processing. They go about processing other orders based on the events they receive and when the event related to your order arrives, it is processed the same way. Now let us see how to implement this in Azure without any serverless components, as in the following figure:

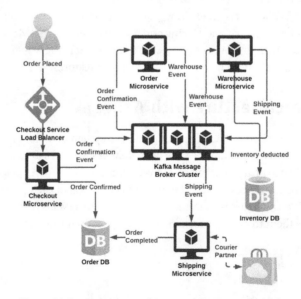

Figure 11.5 – Azure event-driven architecture with VMs

As you can see, all microservices are hosted in VM fleets, and the event routing is done by the **Kafka message broker**, which is also a cluster hosted in VMs. Now let us see how this is transformed into a serverless EDA solution:

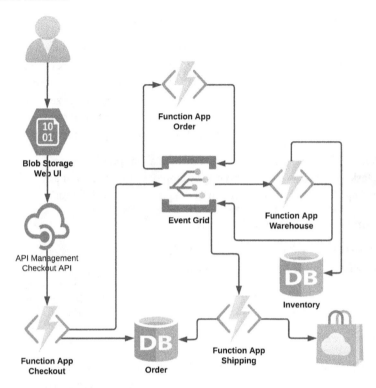

Figure 11.6 – Azure serverless event-driven architecture

As you can see, the frontend is replaced with static web pages in Blob Storage, and dynamic functionalities are provided by an API management platform. The API management platform binds the checkout API endpoint to a **function app** – via the FaaS component of Azure. All other microservices are converted to functions that are triggered on events. The event router in this architecture is **Azure Event Grid**, which is a message broker that creates topics that producers can publish events to. Function apps can be associated with a topic subscription and triggered when an event is published on the topic. For orders and warehouse and shipping events, there will be separate topics. For database access, Azure Functions supports connecting to Azure SQL instances.

Similar to the AWS architecture, this is a minimal reference architecture. The business workflow is a stripped-down version of the real-world scenario and there are a lot more business processes at play in such a transaction. The architecture can also be enhanced with more Azure services. For example, you can add Azure CDN to improve the performance of the static assets in Blob Storage. Some parts of the business logic require workflow management, where you can add an **Azure logic app** instead of a function app.

In the next section, we will look at another use case – business process management. We will use GCP services to build this architecture.

Business process management with GCP

Business process management (**BPM**) is the method that models and executes a business strategy or process. Unlike task management this is the management of the whole process end to end, which involves a number of tasks along with conditions and transformations that make interim decisions. These are repeatable processes, which don't usually change once the process is analyzed and streamlined to match the business needs. As the objective of software development is to translate business needs into software applications most of the time, BPM automation is a crucial part of any application software.

Consider, for example, the simple business process where an employee joins a company. HR has to initiate multiple tasks and processes to onboard the employee into the company in order to enable them to start working. A slimmed-down version of the processes is as follows:

1. Create an employee record in their database
2. Create a payroll account with the payroll vendor
3. Ask IT to create an email ID and other credentials
4. Ask IT to issue a laptop to the employee
5. Ask security to issue a badge
6. Assign a desk – a task for the workplace management team

Now, some of these tasks can be immediately carried out by the HR team themselves, while for a few they have to depend on teams within the company, and sometimes with vendors outside the company. This is a perfect use case for business process automation. Given that this process involves a set of tasks with conditions and transformations, it is a perfect use case for workflow automation.

In a non-serverless world, there are two ways a business process is automated. The first one is to take all the use cases and build one or more microservices to do all these functions. In this case, the workflow orchestration – to process initiating tasks based on their place in the workflow and the conditions that determine whether to initiate a task or not – is done by a microservice that is built on its own. Another way is to take the help of business process management software to do the heavy lifting and create only minimal microservices that are necessary to handle the tasks that this software can't perform on its own. One such piece of software is Camunda (https://camunda.com/), which is a Java-based BPM automation tool that has enterprise offerings as well. Let us look at the architecture to automate employee onboarding with this:

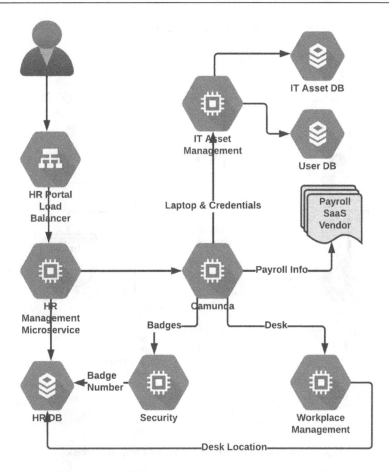

Figure 11.7 – VM based workflow management with GCP

As you can see, Camunda is at the heart of this business process, hosted on GCP VMs. All other microservices are hosted on VMs as well. Camunda sends instructions to various microservices to perform tasks. While the diagram is created to make it clear how the entire business process is orchestrated, it is not necessary for all of the tasks that have to be carried out to be shelled out to microservices. Most BPM software is capable of executing tasks as embedded code or commands within the system. This is possible in serverless as well depending on the capabilities that are supported.

Now, let us translate this to serverless workflow management:

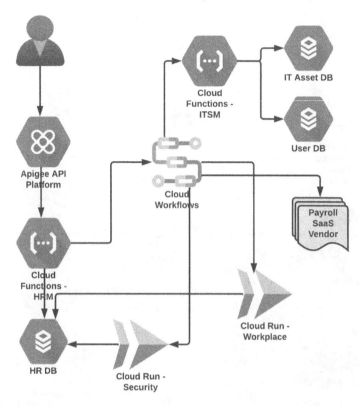

Figure 11.8 – GCP workflow management with serverless

As you can see, each component of the previous design is replaced by a serverless counterpart. For GCP, there are two serverless computing services, Cloud Functions, and Cloud Run. A comparison of these services was given in the *Chapter 5*, in *Part 2*. Essentially, the diagram replaces microservices that need to run longer and need more control and are converted to Cloud Run, while simpler use cases are converted to Cloud Functions. This decision is quite arbitrary based on the assumption of the complexity of these services and could differ in a real-life implementation.

More serverless designs

So far, we have seen three designs across three clouds and compared the serverless and non-serverless solutions. The architecture was specifically designed to demonstrate where serverless services fit it. In the real world though, it is not always like this. There is always a number of combinations of serverless and non-serverless services in the mix. The architecture could even be defined with parts of the design being replaced with different services that fulfill similar needs.

For example, in the non-serverless world, compute instances/virtual machines are often replaced by Kubernetes deployments – especially if the microservice running is of a stateless nature. Similarly, in event-driven architecture, a non-serverless design could incorporate Event Grid instead of managing its own Kafka cluster, which could be operation-intensive. Even in the case of serverless designs, it is not always beneficial to translate all your services to functions, especially if they are vendor-provided or open source solutions. In such cases, these services will remain on VMs or Kubernetes while the remaining infrastructure is transformed into Serverless.

Now that you have an understanding of how serverless design can evolve, we will look into some more serverless designs. Unlike the previous designs, we won't get into comparing them with a non-serverless solution. Rather, we will present smaller, pure serverless solutions that can replace parts of your software architecture. To make it standardized, we will use only AWS technologies in the coming examples.

The webhook pattern

Webhooks are HTTP-based callback functions that are created for responding to events. An ideal webhook should be independently scalable and asynchronous in nature. If a webhook has to run long-running complicated workflows, that defeats the purpose of the design. If you were to run such a long-running process as a response to an external system calling your webhook, you would need to introduce decoupling using a queue, where the webhook is only responsible for gating the external webhook request and placing the data – optionally, after transformation of the request – in a message queue.

Let us look at how we can achieve this using AWS serverless components, as in the following figure:

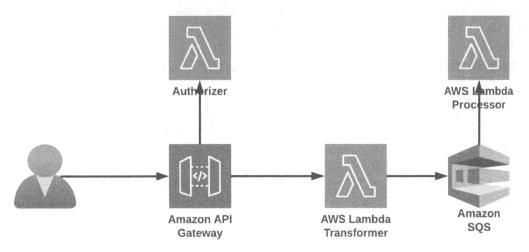

Figure 11.9 – Webhook pattern with AWS

As you can see, the request will be received on an API endpoint supported by API Gateway. If this API needs authentication (most APIs do), then you should have a Lambda authorizer attached to handle that. Once the authentication is passed, the HTTP request is passed on to the Lambda Transformer, which will look into the HTTP request, extract relevant data, create an event out of it, and place it in the SQS queue. Each event in the SQS queue will trigger another Lambda, the processor, which will take the final action on what to do with the webhook request.

Document processing

It is quite common to handle a large number of documents. The usual home for such documents is object storage such as S3. While uploading documents to S3 is a cakewalk, we also need to put effort into extracting information from the documents and storing them for further processing. The following architecture can deal with that:

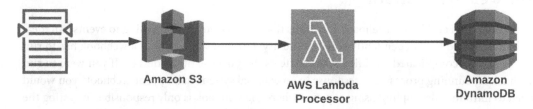

Amazon S3

AWS Lambda
Processor

Amazon
DynamoDB

Figure 11.10 – Document processing with AWS Lambda

As you can see, the document is uploaded to an S3 bucket. Lambda functions can be configured to respond to S3 upload events. So, every time a document is uploaded, this Lambda function runs, processes the document, extracts information from it, and then stores it in a DynamoDB table.

Video processing with the fanout pattern

Video processing is a very common scenario in many businesses, such as services that host educational or entertainment videos by their users. Like the document processing example, videos are also uploaded to an S3 bucket. The upload event will trigger a Lambda function to process the video. When processing a video, it is vital to transcode the video into multiple resolutions to serve different clients, suiting their device properties as well as the network speed. In addition, it is also common to extract thumbnails for previews and to extract audio, which can then also be transcribed. This means a single video processing task spawns multiple tasks to operate on the same video file. This pattern is called the **fanout** pattern. Let us see how we can use serverless to handle this:

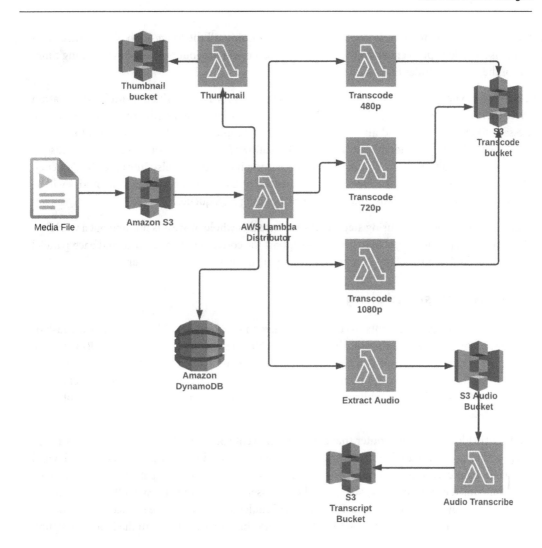

Figure 11.11 – Video processing fanout pattern

As you can see, there is a Lambda called distributor that is triggered when a video is uploaded to a bucket. This then adds the entry into a DynamoDB table with the basic details of the video and a unique ID. Then it goes on to spawn a number of other Lambda functions for jobs such as transcoding, thumbnail and audio extraction, and so on. Each of these tasks saves the resultant audio or video in corresponding S3 buckets. In the case of audio, when it is pushed to an S3 bucket, which triggers another Lambda that will transcribe the audio, thus producing the transcript of the video and storing it in the same DynamoDB database.

Now that you have understood how fanout works, let me point out a flaw in this design. All the tasks that the distributor Lambda spawned need to be watched to ensure they are successfully completed.

If the distributor were to do it, it would have to keep running till the longest job was finished. This could cause a serious problem since the distributor itself could timeout or cause long running times if one of the child Lambdas takes a lot of time.

The right way to deal with this is to use a service built exactly for fanout. That is **Simple Notification Service (SNS)**. SNS provides a publisher-subscriber model, where the distributor Lambda can drop a message to an SNS topic, and all the subscribers to that topic (the **child Lambdas**) get the same message with the information about the video. This allows for better fanout and retries in case of message delivery failures. If the volume of messages is too high, you can also introduce SQS queues, where the queue will subscribe to the SNS topic instead of Lambdas, and then the Lambdas will be triggered with the messages delivered to the corresponding SQS queue.

Another way to handle this is using **step functions** since this whole process is nothing but a workflow with a number of parallel steps. Step functions support the construct of parallelism and each parallel step can be a child Lambda. Step functions also support their own error handling.

Serverless job scheduling

There are often time requirements to run jobs at a given point in time, either as a one-time task or repeatedly. In the non-serverless world, this is usually achieved either by using the OS's built-in crontab scheduler or by continuously running a service/daemon that consults a database of schedules at defined intervals and launches the corresponding tasks. Different clouds solve this problem in different ways. For GCP, it is Cloud Scheduler, while in Azure, Logic Apps deals with schedules. In the case of AWS, it is **AWS EventBridge**.

AWS EventBridge is an event router that connects an event source with a target. There are various sources and targets supported by EventBridge. They are connected by filtering events with rules and then invoking the corresponding target. In our case, we are interested in its capability to run scheduled events. Scheduled events are generated periodically based on a `cron` expression that we configure and can invoke any supported targets. Typically, a scheduled job is a compute service, most probably a Lambda function. The design is quite simple, yet powerful and is as shown in the following figure:

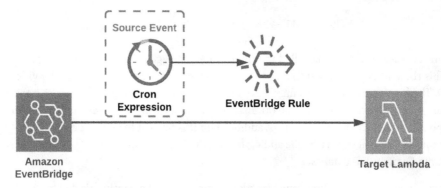

Figure 11.12 – Serverless job scheduling with AWS EventBridge

There are a lot more patterns similar to this that you can achieve with serverless, but what we have covered so far should give you enough foundational design ideas to work on any cloud. We will conclude this chapter with some of the serverless design principles covered in the AWS Well-Architected Framework.

Serverless applications in the Well-Architected Framework

The AWS Well-Architected Framework has a guiding document called the **Serverless Applications Lens**. This document covers common serverless design scenarios and lays out best practices. The core design principles are as follows:

- **Singular**: Functions should be single-purpose and simple enough to satisfy the one case they serve.

- **Concurrency is the key**: Any design decisions you take should be based on how many concurrent invocations the use case would need, as serverless is always tuned for concurrency.

- **Stateless**: FaaS is ephemeral in nature and doesn't store anything in its own environment. It should share nothing and you should assume that its execution state and underlying storage are perishable. If durability is part of the use case, use persistent remote storage.

- **Orchestrate with state machines**: While it is possible to chain Lambda functions to orchestrate complex workflows, it will defeat the very foundational principles of serverless design by creating a tightly coupled monolith. Hence, it is always recommended to use a cloud workflow engine for complex orchestrations.

- **Prefer event-driven design**: Most cloud services emit events and any logic that is to be modeled around these changes should use asynchronous event-driven design.

- **Design for failures and duplicates**: Events and requests can be corrupted and fail to process. It is advised to incorporate appropriate guardrails against these failures. Similarly, in EDA, events can be delivered more than once, and the logic of your design should take care of the deduplication.

Summary

This chapter was about design in general and serverless design in particular. We started off with OOP design patterns as that serves as the foundation of designs for developers. We briefly covered architectural and integration patterns to complete that foundation. Then we described the relevance of the Well-Architected Framework and how it helps in cloud architecture design. Finally, we covered a number of serverless architectural patterns that are used to solve common business problems in the cloud.

This concludes our chapter, and as the last chapter of *Part 3*, it is also where we conclude this book. In this book, we have covered a lot of ground around the serverless theme. The first part laid the groundwork by discussing the foundations of cloud computing, the general design principles of various serverless cloud services, and their relevance. In *Part 2*, we covered serverless technologies from various vendors, and how to configure and run them. We wound up each chapter with a small project, to sum up what you learned in the chapter. In the third part of the book, we covered general best practices for development and security. We also introduced a number of vendor-agnostic frameworks to manage your serverless applications and infrastructure.

The idea of writing this book was to open up the world of serverless to readers. It was not about diving deep into one technology but making you understand that irrespective of the technology stack you are bound with, you can always derive goodness out of serverless. If this book has been helpful in taking you in that direction, then it has achieved its mission.

Thank you!

Index

Symbols

A

www.packtpub.com

Subscribe to our online digital library for full access to over 7,000 books and videos, as well as industry leading tools to help you plan your personal development and advance your career. For more information, please visit our website.

Why subscribe?

- Spend less time learning and more time coding with practical eBooks and Videos from over 4,000 industry professionals

- Improve your learning with Skill Plans built especially for you

- Get a free eBook or video every month

- Fully searchable for easy access to vital information

- Copy and paste, print, and bookmark content

Did you know that Packt offers eBook versions of every book published, with PDF and ePub files available? You can upgrade to the eBook version at packtpub.com and as a print book customer, you are entitled to a discount on the eBook copy. Get in touch with us at customercare@packtpub.com for more details.

At www.packtpub.com, you can also read a collection of free technical articles, sign up for a range of free newsletters, and receive exclusive discounts and offers on Packt books and eBooks.

Other Books You May Enjoy

If you enjoyed this book, you may be interested in these other books by Packt:

Go for DevOps

John Doak, David Justice

ISBN: 978-1-80181-889-6

- Understand the basic structure of the Go language to begin your DevOps journey
- Interact with filesystems to read or stream data
- Communicate with remote services via REST and gRPC
- Explore writing tools that can be used in the DevOps environment
- Develop command-line operational software in Go
- Work with popular frameworks to deploy production software
- Create GitHub actions that streamline your CI/CD process
- Write a ChatOps application with Slack to simplify production visibility

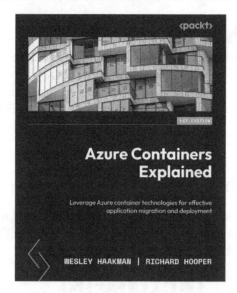

Azure Containers Explained

Wesley Haakman, Richard Hooper

ISBN: 978-1-80323-105-1

- Make the best-suited architectural choices to meet your business and application needs
- Understand the migration paths between different Azure Container services
- Deploy containerized applications on Azure to multiple technologies
- Know when to use Azure Container Apps versus Azure Kubernetes Service
- Find out how to add features to an AKS cluster
- Investigate the containers on Azure Web apps and Functions apps
- Discover ways to improve your current architecture without starting again
- Explore the financial implications of using Azure container services

Packt is searching for authors like you

If you're interested in becoming an author for Packt, please visit authors.packtpub.com and apply today. We have worked with thousands of developers and tech professionals, just like you, to help them share their insight with the global tech community. You can make a general application, apply for a specific hot topic that we are recruiting an author for, or submit your own idea.

Share Your Thoughts

Now you've finished *Architecting Cloud-Native Serverless Solutions*, we'd love to hear your thoughts! Scan the QR code below to go straight to the Amazon review page for this book and share your feedback or leave a review on the site that you purchased it from.

https://packt.link/r/1803230088

Your review is important to us and the tech community and will help us make sure we're delivering excellent quality content.

Download a free PDF copy of this book

Thanks for purchasing this book!

Do you like to read on the go but are unable to carry your print books everywhere?

Is your eBook purchase not compatible with the device of your choice?

Don't worry, now with every Packt book you get a DRM-free PDF version of that book at no cost.

Read anywhere, any place, on any device. Search, copy, and paste code from your favorite technical books directly into your application.

The perks don't stop there, you can get exclusive access to discounts, newsletters, and great free content in your inbox daily

Follow these simple steps to get the benefits:

1. Scan the QR code or visit the link below

https://packt.link/free-ebook/9781803230085

2. Submit your proof of purchase
3. That's it! We'll send your free PDF and other benefits to your email directly